Empires of Speed

Time and the Acceleration of Politics and Society

By

Robert Hassan

BRILL

LEIDEN • BOSTON
2009

Cover illustration: Photo by Éole—http://flickr.com/eole.

This book is printed on acid-free paper.

Library of Congress Cataloging-in-Publication Data

Hassan, Robert, 1959–
Empires of speed : time and the acceleration of politics and society / by Robert Hassan.
p. cm.
Includes bibliographical references and index.
ISBN 978-90-04-17590-7 (hardback : alk. paper) 1. Time—Sociological aspects. 2. Information society. 3. Information technology—Social aspects. 4. Computers and civilization. I. Title.

HM656.H38 2009
303.48'3301—dc22

2009012416

ISSN 1873-7463
ISBN 978 90 04 17590 7

PRINTED IN THE NETHERLANDS

For Barbara Adam, Mentor and Friend, who inspired this book.
To my Family, Kate, Theo and Camille without whose support it could not have been written.

Prosperity never before imagined, power never yet wielded by man, speed never reached by anything but a meteor, had made the world irritable, nervous, querulous, unreasonable and afraid.

Henry James, *The Education of Henry James*, 1918

Time has been a most effective colonizing tool.

Barbara Adam, *Time*, 2004

CONTENTS

INTRODUCTION

End of Empire?

In one sense the word 'empire' has always had a singularly Victorian ring to it. Related to the word 'imperial', in Britain it originally connoted (and was painstakingly intended to connote) the grand national project of bringing civilization to the less fortunate peoples of India or Africa or other non-European regions. And so when I survey the literature of the time and think of what it must have conjured up in the minds of the British of the 19th and early 20th centuries, I see Great Exhibitions of stupendous machinery and their marvelous products; I see earnest missionaries going about their good works in the face of some oddly recalcitrant subjects; and I see the British middle- and upper-classes bringing bureaucracy and administration to these disordered regions in the shape of English Laws, English values and Enlightenment Reason.

Of course this is idealization. The reality was much less spiritually or culturally fulfilling for most people. It was a model propagated in the media of the time to spread a veneer over a project of plunder for the factories and coffers of industrial capitalism; the willed extinction of peoples and cultures, and the physical subjugation of those within these territories who would think to resist.

In the early 20th century there began to be critiques developed from Marxist or anarchist or liberal traditions that would expose the less-than-ideal reality. Gradually the word 'empire' began to be tinged with other meanings. The term 'colonialism' like 'imperialism' became synonymous not only with empire, but also with the perpetration of robbery, repression and ruthlessness on an enormous scale. And as the century went on, the contradictions became more pronounced. For example, at the political level, arch-British Empire defender Winston Churchill was able in the 1930s to warn against the 'mortal danger' of a Nazi empire forming in Central Europe, whilst at the same time clinging unselfconsciously to one that was being controlled, albeit with growing difficulty, from the Foreign Office in Whitehall. By World War Two's end, and through to the 1960s the theory of empire had been upended by successive United States governments who looked to support those independence movements that were oriented towards the Washington worldview. 'Empire' had become a nasty word, an anachronism and an affront to liberal democratic values that now seemed to have its natural home, as de Tocqueville had once predicted, in the New World across the Atlantic. The British, French, Portuguese, Dutch and Belgian 'possessions' were wrested forcibly or, less often, peaceably, from their owners by new nationalist movements that felt they had the winds of history at last filling their sails.

The idea of empire became truly contemptible by the late1960s,—especially when the US was seen in some quarters to be hypocritically shoring it up in its entanglement in Vietnam. The final retreat from this part of Southeast Asia in 1975 seemed to put an end, forever, to this particularly Victorian project. The term 'empire' had by now become a weapon of ideological invective by the liberal democracies that had invented it in the first place. The cycle seemed to have turned fully by 1982 when U.S. President Ronald Reagan referred to the doomed Soviet Union as an 'evil empire'. The historical irony, of course, is that over this period an 'American Empire' based on the 'soft power' of its culture backed by corporate capitalism continued to grow unchecked in the form of a myriad of cultural commodities, from Hollywood to Coca-Cola, and from Nike to iPods.

Much of this is well known. The purpose of this introduction, however, and the purpose of the rest of the book in fact, is to think anew, not about the 'idea' of empire, but to look at empire in a new way. What this perspective will do is to show that the *political and economic process of empire* has never gone away. I will argue that beginning before the industrial revolution and then in a transformed way with the rise of the information technology revolution, two *temporal* empires have ruled over the inhabitants of our modern (and now postmodern) economies and societies: one based upon the clock and the other upon a ubiquitous information network that comprises the 'network society' and what I have elsewhere called 'network time' (Hassan and Purser, 2007). These have been *invisible empires* because our Enlightenment-based form of time was born as an abstract concept that developed and spread to become a social, industrial and political process, through what Herbert Reid saw as the technological 'conditioning' of our human consciousness (1973:201).

The *spatial* empire of the British or the French was obvious—as obvious to the map-drawers in London and Paris as it was to resisters in their jails in Bombay or Algiers. The temporal empire, by contrast, was not so obvious—even less apparent than the subtle U.S. empire built on the 'soft power' of New World cultural commodities. The temporal empire exists in our heads through the 'conditioning' that Reid speaks of. It has led us to imagine (or simply take for granted through being born into it) that the clock represented what time is—a 'linear, measurable, predictable time' (Castells, 1996:433) whereas I argue that it is in large measure a reflection of the dominant values of Newtonian-Enlightenment science and technology. Moreover, this form of time reckoning is being displaced by a new relationship to time and to speed, a transcendence that constitutes the rise of another temporal empire based upon a computer-based 'real-time' that is undermining the dominance of the clock that has conditioned economy and society for over 250 years (Hassan, 2003).

These empires of time and of speed become more apparent if we see time and space as a process of what Thrift and May (2001) call 'timespace'. That is to say, as processes that both interpenetrate and mutually influence and, centrally, are *socially created*. In this view time and space do not exist as 'absolutes' existing outside of us, as Isaac Newton would have it, but instead we as social creatures *create time* through social contexts and through the practice of culture in an environment of temporal diversity that has dimensions of biological time, cosmic time, social time and so on. In short the temporal empires I will describe *are* empires because they are technologically based forms of time (clock and computer) that dominate other forms of time reckoning and occludes other ways of thinking about time.

The end of politics?

Whether viewed in its traditional or its temporal sense, empire is still underpinned by a politics that forms the basis of its social organization. Consequently this is a book that is fundamentally about politics. More specifically the book is concerned with the nature, role and function of liberal democratic politics in the context of a high-speed networked society. Liberal democracies come in many forms. France is different from the New Zealand, for example, in that the former has a formal Constitution and the latter does not. Another difference is that the United Kingdom holds general elections every five years and in Australia it occurs every three. There are more fundamental differences too in, say, what is proclaimed as democracy in Russia, and how it functions in Spain. At the level of detail, then, quite major differences abound. However, to be able to be applicable at the level of the global, my idea of what constitutes a liberal democracy is necessarily broad so as to encompass most countries in the world. And so in the context of this argument liberal democracy is defined as being characterized by a representative system of government that is subject to regular elections, it subscribes (at least in theory) to free speech and free association, and it extends the vote to the whole of the population (again, at least in theory). My argument is that as presently defined and constituted in the context of a high-speed network society, liberal democracy is unable to deliver upon its historic responsibility to be a 'government of the people, by the people, for the people' as Abraham Lincoln put it.

The form of political philosophy that has dominated in the English-speaking world since the 1970s—and has made significant inroads into the economy, culture and society of the rest of the world—has been termed *neoliberalism.* It is a term that connotes a transformation to a more market-oriented world, one based on individualism, on consumerism, on a leading role given to 'market mechanisms' and to information technologies, and on the supposed efficiencies and productivities that this nexus would bring. The prefix *neo* does indeed connote a change in the role and effect of liberal democracy today, but in my argument it also suggests a new temporal dimension, a new empire of speed—one that casts the assumptions regarding the benefits of neoliberalism into a very harsh light.

It is sometimes claimed that the 'rough and tumble' of liberal democratic polities and societies is evidence in itself of a certain positive dynamism, where checks and balances are called upon to perform their intended function, and in so do doing 'show' that the system 'works', at least on a heuristically evolving basis. Neoliberalism is supposedly no different in this respect. The trial and guilty findings of Kenneth Lay and Jeff Skilling of Enron Corporation in 2006, for example, is often cited as evidence that capitalism is, in the final analysis, answerable to the law, and through it, to the government and the people. This view of neoliberal democracy as the highest expression of our political development was the centerpiece of Francis Fukuyama's thesis in the *End of History*, where 'history' as the working out of the major contradictions in economy and society had finally run its course. All that was left to do, according to Fukuyama, was to work through the details, and iron out the creases in the way that federal district court Judge Sim Lake might have been seen to be doing when sentencing Lay and Skilling to lengthy jail terms. However, an honest accounting of the promise of liberal democracy, and especially a neoliberal democracy, when set against the reality of its effects, makes the democratic ledger look, well, disastrous. We will discuss the many negative social and political effects of neoliberalism in the following chapters, but let me preface that with a quotation from the theorist of deconstruction, Jacques Derrida, who was prompted to respond to Fukuyama's insouciance in his 1993 book *Specters of Marx* where he writes that:

> ...it must be cried out, at a time when some have the audacity to neo-evangelize in the name of the ideal of a liberal democracy that has finally realized itself as the ideal of human history: never have violence, inequality, exclusion, famine, and thus economic oppression affected as many human beings in the history of the earth and of humanity. Instead of singing the advent of the ideal of liberal democracy and of the capitalist market in the euphoria of the end of history, instead of celebrating the 'end of ideologies' and the end of the great emancipatory discourses, let us never neglect this obvious macroscopic fact, made up of innumerable singular sites of suffering: no degree of progress allows one to ignore that never before, in absolute figures, have so many men, women and children been subjugated, starved or exterminated on the earth (p. 13).

From this critical perspective, an argument is developed that claims that our neoliberal society and networked economy functions upon an open-ended continuum of speed, and one that bears no small responsibility for the 'sites of suffering' that Derrida alerts us to. It is a tempo driven by a freed-up capitalism, an evolved economic system that operates on the basis of 'speed equals profit'. Capitalism achieves this economic acceleration by suffusing information and communication technologies (ICTs) to a ubiquity across every register of life, whilst at the same time exacting an immense social and cultural price.

A theory of temporality and democracy

The connection between temporality and democracy is of course central to my thesis, and is something we shall explore in some depth as the argument develops. However, let me sketch out the contours of this connection so as to preface that discussion.

Regis Debray has argued that the dominating ideas of a society are underpinned by the 'material forms and processes through which its ideas are transmitted', that is to say, through the 'communication networks that enable thought to have social existence' (2007:5). The ideas of modern liberal democracy evolved and grew to inhere within the specific 'material forms' and the 'communication networks' of an in tandem emerging industrial capitalism. Capitalism did not invent the clock, but its indispensable use as a powerful rationalizing and instrumentalizing force for capital accumulation meant that it soon settled at the very center of the industrializing process. As the clock began to condition the consciousness of industrializing culture and society, then so too did it inflect and shape the politics of liberal democracy through which capitalism was able to grow. This crucial nexus between capitalism, clock time and liberal democracy was able to function more-or-less effectively, as we shall see, for over two hundred years.

However, with the coming of an ICT driven neoliberalism—more a form of capitalism than a form of politics—the significance of the clock begins to wane. The relative demise of the importance of the clock increases as we become increasingly networked. We begin to inhabit a 'virtual' space and time with every new (and faster) digital application, product or program that comes to the market to make our lives supposedly more efficient and in line with the new imperatives of a neoliberal economy. Historically, liberal democracy as a form of governance had a social responsibility to plan and to organize and to be future-focused at least as far as its legislative arm was concerned. Today, it is unable to synchronize with a fast-paced, instrumentally oriented economic neoliberalism that is focused far more on the short-term horizon, and on doing things ever more quickly to satisfy profit-driven expectations. The result, in very simplistic terms, is that liberal democracy does not now work, indeed cannot work in a neoliberal context that valorizes speed in the search for profit, and raises efficient and productive profit-seeking to be the central criterion of what it is to be human.

To understand more clearly what this means the book brings to the fore—makes explicit—a theory of temporality. It is a theory that illustrates the profound importance that the clock and the network has for our society and for the constitution of our political system. In so doing, the book will help to close a gap that constitutes a relatively under-theorized space in the social sciences and in accounts of politics and the information society more generally. A key concern, then, is that whilst many would agree that life seems to be getting faster, and that ICTs have a role in this—what does this actually mean, and more pertinently, what are the consequences for economy, society and politics?

The networked society

That we live in a networked society is now a given. There are reminders of it everywhere. It is in evidence in every street where people talk into cell phones, or listen to MP3 players; we see it on every factory floor or office, where the flicker of the computer screen and click of the keyboard are the pervasive sight and sound; and we see it in the home where the networked computer takes on the focus that used to be enjoyed by the television set—indeed the computer has colonized the home and has incorporated television into itself to become yet another function of networked life. But what kind of society is this? Manuel Castells sees the networked society as one of 'flows' that is 'the expression of the processes dominating our economic, political, and symbolic life' (1996:412). Such flows suggest constant movement and change, but to say that these are change-filled times is to indulge straightaway in cliché. Now and then, however, resort to the hackneyed concept can be a useful entrée into fresher and more interesting ideas. So let us quickly dispense with the formulaic claim: *in these early years of the 21st century, polities, economies, cultures and societies across the world are changing with a rapidity that is unprecedented.* This is a truism, but it is one that this book shall dig deeply into to reveal some interesting (and alarming) dynamics at work. To make the job easier, we can begin by separating out two major implied ideas within this well-worn sentence. One is the all-encompassing nature of the claim—its *globality.* The other idea is contained in the words 'rapidity' and 'unprecedented'. Let us briefly look at these in their turn.

Globality—or globalization—is a term we are well used to by now. It connotes, mainly, economic change, with the world becoming 'networked' and 'digital', and with the Internet and new media technologies driving the process towards one world—or many connected world's—depending upon whom you read (Lechner and Boli, 2007). The consensus point is clear, though: for reasons I shall go into below, the world has become a singular, interconnected place where major changes tend to have effects and implications for nearly everyone, be it the price of oil, or shares, a meltdown in Wall Street, or the mutation of an influenza virus in pigs or birds, the effects of climate change, the introduction of a more powerful microchip, and so on.

The terms 'rapidity' and 'unprecedented' are here also deeply connected to globalization. These are expressions of its particular tempo, its scope and its direction. To take the first word: we have never experienced such a world were 'rapidity'—*speed—is* at the very core of our collective and individual existence. As for the second, the transformation to a speed-filled world—filled because it reaches into every realm of the social, economic and cultural, into our very consciousness—is indeed 'unprecedented'. And globalization, with its rapid pace and its revolutionary nature, is profoundly political, of course. Politics and politicians have created the context in which neoliberal globalization has been allowed to evolve. But the interpretation of the transformation that this book seeks to articulate, goes well beyond the traditional theorizing of politics. In fact it is an interpretation that signals the *death of the politics of liberal democracy*—with nothing (or at least nothing democratic) to replace it.

Neoliberalism, or, when democracy succumbs to the economy

Let us take these introductory remarks a bit more gradually, and contemplate a few questions that relate to what kind of politics is being discussed in this book and in what context. One question is: why do liberal democratic political institutions seem not to meet the needs of the majority of people any longer? Increasingly at every election, in almost every liberal democratic country, millions shun the voting booth, and are uninterested in the drone of mainstream political discourse. People consider the institutions of democracy in the parliament, or congress or wherever, as unresponsive to their concerns, and view its workings as being disconnected from the realities that shape their everyday life. It is a response that for some stems from a ground-down apathy, for others it comes from a growing cynicism, and for others still, as we shall later see, it emerges from a preoccupation with other forms of politics in other political realms.

Within all these attitudes, a loss of faith is starkly reflected in our perceptions of individual politicians—in that constantly smiling, media pervasive and monotonously besuited and overwhelmingly male physical embodiment of the liberal democratic ideal. For example in 2005 the Gallup organization surveyed a large-scale representative sample of some 50,000 people across 68 countries on their attitudes towards a range of social issues. In terms of their levels of trustworthiness, the political class managed to achieve a derisory 13 per cent approval rating (Whitaker, 2005:7).

This globalized negativity is not directed towards liberal democracy *per se*, but to its mutated form and its buried ideals in the shape of *neoliberalism*, a political, economic, social (and technological) worldview that emerged with tremendous force in the late1970s. The rhetoric of neoliberalism, especially in the 1970s and 1980s, when it strove hardest for intellectual and political legitimacy, argued that it was more democratic than the allegedly 'statist' social democracy experiment of the post World War II years. Neoliberalism was thus supposedly a return to a less complicated time of Adam Smith and Alexis de Tocqueville, to a time when the 'laws' of supply and demand were first 'discovered' and promoted, and to a time when the connections between democracy and the 'free market' were seen by many of the *philosophes* of the Enlightenment to be as obvious as they were natural.

Today, as an idea and, much more importantly, as an *ideology*, the tenets of neoliberal politics dominate the political landscape. Moreover, neoliberalism pervades a global media that serves as its primary ideological vector. Its stock phrases weave through the speeches of politicians and its mathematical abstraction of real world processes inform the similarly abstract worldviews of powerful economists in the World Bank and International Monetary Fund (Klein, 2007). Notwithstanding the global power of the neoliberal outlook on how the world actually is, possibly never before have the abstract idea and the concrete reality clashed with such widespread and devastating effect. On one hand neoliberalism promotes the idea of the market and free trade and the much-reduced role of the state in economic and social affairs as the only sure road to progress and prosperity. Implicated in this dogma is the notion of democracy. For example, according to Friedrich von Hayek, one of the formative influences on neoliberal thinking, the economic freedom of the market would ensure that 'coercion of some by others is reduced as much as possible' and this, he goes to say in his *The Constitution of Liberty*, is a form of freedom in its 'original meaning of the word' (1978:11). Moreover, and making the vital connection for us, von Hayek points out that a market-based freedom also embraces a 'civil liberty' and a 'political liberty' as well (p. 12). Accordingly, neoliberal societies are said to be more democratic because *free* markets and *free* trade suggest individual and institutional autonomy, a freedom to choose how to live and how to work.

To feel the weight of the other—and much heavier—hand of neoliberal reality it is necessary at this point to build on the assertions of

Derrida that were quoted earlier, and refer to the reported findings of those who occasionally get out from their offices to see what happens when theory meets with practice. In 2005 the UN published its *Report on the World Social Situation.* This is a regular publication and is backed by respected academics and NGO workers, and is promoted by the UN Secretary General in the world's media. However the 2005 edition, like its predecessors, hardly made a blip on the radar of the global mediascape. At best media coverage amounted to a column or two in the major newspapers and a few seconds airtime on radio and television. The 2005 *Report* was subtitled 'The Inequality Predicament'. The problem of inequality is usually seen as an economic one, but it has obvious political ramifications too. Inequality, as von Hayek himself would undoubtedly argue, is a political 'unfreedom' that stems from economic exploitation and from marginalization from the main sources of economic activity. The *Report*, however, is devastatingly clear on the trends concerning global inequality over the last twenty years of neoliberal domination of the global economy. It notes with concern 'the widening gap between skilled and unskilled workers [and] the growing disparities in health, education and opportunities for social and political participation' (2005:ii).

Another turn-off for the voter is the hypocrisy that many people smell emanating from the practice of neoliberal democracy. As Michael Hardt and Antonio Negri point out in their book *Multitude*, neoliberalism and its ideology of the freedoms that free markets supposedly bring, is based upon an immense falsehood. Neoliberalism does not function through vigorous competition and through the unregulated market mechanism of supply and demand. As the yearly convocation at Davos in Switzerland shows, the reality of neoliberalism is that 'the economic, political and bureaucratic elites of the world need to work together in constant relation' in the pursuit of profit and the efficiency and flexibility of their operations (Hardt and Negri, 2005:167). This is how neoliberalism as practiced by the elites functions: a top-level strategizing and cooperating to ensure their mutual (mainly shareholder) benefit to the detriment of the vast mass of humanity—not to mention the detriment to the natural environment that many neoliberal 'business friendly' policies bring. To paraphrase Zygmunt Bauman, this new order functions in the context of a new temporality that creates new forms of polarization (1998:18).

The double standard involved in corporations exposing the lives of workers to the traumas of free market flexibility and not exposing themselves to it in anything like such a radical degree, has been in evidence since the dawn of the neoliberal age in the 1970s. We see one side of the duplicity wherever workers protest at being forced to respond to ever-changing market conditions, usually to the detriment of pay, working conditions and job security. And we see its obverse, daily, in mainstream media, in business and in government, and in increasingly powerful right-wing think tanks, exhorting workers to be efficient and flexible no matter the circumstances, because their very livelihoods depend on it. The same logic works at the level of international trade where there is another set of laws for global corporations who can complain to the World Trade Organization (WTO) of 'unfair competition' from subsidized Nicaraguan banana farmers or Kenyan coffee-growers. And if this fails then they can come to a cozy inter-corporate agreement the next time Davos comes around.

The trenchant political message from all of this, one that millions are easily able to work out for themselves, is that liberal democracy, or more accurately *neo*liberal democracy, is fundamentally biased against people at the grassroots level, at the level where democracy is supposed to spring from. Politics today fails not principally because individual politicians are malign and corrupt and cynical and interested only in the business of business—although more than a few undoubtedly are: politics fails because the institutions of democracy, the institutions of parliament and congress and of local government assemblies, cannot democratically function because of the effects of what William Scheuerman calls the 'high speed economy' where there exists a 'mismatch between the time horizons' of the institutions of liberal democracy and the 'built-in' accelerative drive of the capitalist system (Scheuerman, 2003).

The value of the temporal perspective

How this temporal 'mismatch' or 'asynchronicity' or 'break' between the political system and the economic system came about is a key theoretical concern of this book. By concentrating on the temporal framework of analysis, the nature of the failure of liberal democracy will be highlighted. The overall thesis will thus provide a fuller picture of what constituted liberal democratic politics in its modernist phase, as well as explain the motive force behind neoliberal-influenced politics in their present 'postmodern' articulation. Moreover, the temporal perspective ensures that many of the social and economic 'pathologies' of our postmodern world, that are analyzed in Chapter Four, are seen not as disconnected phenomena—as tends to be the case with much traditional social science analysis—but are viewed as part of a logic that stems from the particular evolution of the relationship between capitalism, liberal democracy and the application of Enlightenment-based science and technology. Scheuerman's work on the temporal dimension to the political process is pioneering stuff. Nevertheless he works principally within the realm of political science and so for him the political process tends to be the beginning and the end of the story. Moreover his temporal dimension is focused primarily on the 'high-speed' society as an unexplained given. In this book however, politics is the end of the story, not its beginning. And as a social theorist of time I want to begin the book by offering a deeper understanding of theories of time in the social sphere: why, for example, do we live in a 'high-speed' society? Why does is have to get faster, and what are the effects of what Hartmut Rosa (2003) has termed 'social acceleration'?

By being 'time sensitive' to the question of 'what is liberal democracy?' the scope of this book is broadened out considerably. By 'temporalizing' the dynamics and the social effects of the interactions between the political institutions of liberal democracy and the processes of capitalist industrialization that began to combine in the 18th century, the theoretical lens switches to wide-angle. It makes apparent the birth, evolution and periodization of two immense *empires of time* that are based upon our historical relationship with temporality as mediated through politics, through the logic of capitalism, and through the dogmas of science and technology.

We begin to approach the nucleus of what guides much of the investigation. Another question: What do you get when you foreground the temporal perspective in social science analysis? That is to say, what do you uncover or make more apparent when the time factor is made explicit? It is a fair question. And it is one that was rhetorically posed by Hebert Reid in a long-ago pre-neoliberal and pre-network time in the early 1970s. He tried to expose a lacunae in political theory, one that has not been properly addressed since, when he noted that 'We have hardly begun to explore either the intimate or large-scale ways in which the dominant time-orientation has conditioned the public realm and agenda in terms of which our political life is supposedly conducted' (1973:202).

By way of a preliminary answer, let us consider the quotation that was contained in an email I received recently from a listserv that is concerned with theoretical issues around computer technology on the general level and with computer networks more specifically. The quotation was from Sam Walton, the founder of another empire, that of Wal-Mart retailing, and it reads: 'everyone thinks Wal-Mart's success is because of economies of scale. The real reason is that we replace inventory with information'. The context of the email was to rather blandly show the significance of computer-based information for businesses in today's networked economy. It is a point emphasized (and made rather more interesting) when it is considered that in the USA Wal-Mart's computing power in the 1990s was second only to the Pentagon's (Time, 1992).

Consider Walton's quote from the perspective of temporality. At a surface level the 'inventory-to-information' tale of Wal-Mart may be seen as symptomatic of a more general change in the way companies became more efficient. It would indicate, fairly simply, that Wal-Mart led the way in the US in the replication of the just-in-time production methods that was pioneered by Japanese automaker Toyota in the 1960s. However at a deeper level of temporal understanding, the Wal-Mart story feeds in to the time-influenced work of Jeremy Rifkin and his theory of the 'weightless economy' (2001:30–55). Building on themes he explored in his 1987 book *Time Wars*, Rifkin deftly shows how in the brave new world of deindustrialization and digital networks, that new sources of wealth and profit were to be found in knowledge creation, information and ideas—not in weighty physical assets such as factories and warehouses that are fixed in space. A science and technology writer and sometime lecturer in corporate management, Rifkin's analysis is nonetheless a temporalized form of Marxist political economy. He shows clearly that profit is derived from speed and mobility. This was a tenet of Marx's own writings, but it was a lesson that had been forgotten in the long years of Fordist ascendancy where fixed assets and economies of scale and the 'managed economy' where seen to be the optimal source of profit.

By making explicit the importance of speed that Walton's quote implies, we can go to a deeper (and wider) temporal perspective on what makes modern (and postmodern) economies tick—and what makes them change. This takes us to David Harvey's magisterial work, his 1989 *The Condition of Postmodernity*. For Harvey the intrinsic need for capitalism to spatially expand was the dynamic behind the transition from the Fordist mode of production to the post-Fordist regime of 'flexible accumulation'. Geographic space was 'annihilated' by time through the revolutionary advances in ICTs, a revolution that gave information its present-day status of primary importance that Sam Walton understood so well. This economic necessity produced the 'acceleration of physical processes' (Harvey, 1989:22) that were taking their temporal cues from the open-ended and constantly increasing speed-dynamic of computerization.

But even at this level of temporal acuity, we do not yet get to the nub of what time is. For that we need a general social theory of time, and we need to place it at the core of our political economy, our political science, and our equally important analyses of media, cultural and social processes. As noted previously time theory has been a marginal part of the social sciences. However, this has been changing in recent years, within sociology especially, and not least as a response to the increasingly palpable 'problem' of time in everyday life—expressed in the so-called 'time-bind' in work and in leisure (Rifkin, 1987; Schor, 1993; Hochschild, 1997).

The general social theory of time that unfolds in these pages draws upon the philosophers of time such as Edmund Husserl and Henri Bergson, who tried—in a conscious break with the dominant clock-based perceptions—to uncover the experiential 'essence' of time and duration. Moreover, the work builds upon the ideas of leading contemporary social theorists such as Barbara Adam, Helga Nowotny, Ida Sabelis and Hartmut Rosa, who all place temporality at the centre of their analysis. Through the works of these and others *Empires of Speed* develops a general theory of time that allows us to understand why we find it so difficult to think about time other than in the context of the clock. The book renders the nature of time more graspable. For example, it allows us to realize, indeed to be reminded, because it a fact that we intuitively know, that time is in everything—Nowotny uses the term (*eigenzeit*). This can be illustrated by recognizing that as humans we are bounded by our time-life span, but within this duration there are many other temporalities. There are the diurnal rhythms of sleep, or the psychologically based memories of the past, our sense impressions of the present, or even our mental projections of the future. These temporalities are an innate part of our human make-up. It gets even more diverse as we dig deeper into the details of our biology:

there is the (more-or-less) nine-month gestation period of the egg to baby, or the differing times it takes a child's tooth to grow, or for our hair to sprout, or fingernails to lengthen, or the changeable rhythms of a heartbeat, and so on. Temporal diversity exists 'outside' us too. The grass or pavement that we walk on, for example, have their own times—the relatively short time that it took the grass to grow, or the different times 'folded' into the pavement concrete through the ages of the quartz, rock and sand or whatever natural substance it may have been comprised of (Latour, 2002:249).

To this 'intuitive' recognition will be attached a more rigorous theoretical framework of temporality. Time is a dynamic social and biological process, and the temporalities that exist within us and which permeate our built and natural environments come together (intersect) to comprise what Barbara Adam calls 'timescapes' (1998). From the perspective of the 'timescape', as we shall later discover, we can appreciate more fully the role of technology in what is an increasingly complex temporal process. As the processes of modernity and industrialization developed and became more sophisticated and widespread, then so too did technologies and productive techniques and patternings became 'entimed' by the growing power of the clock, a technology that for Lewis Mumford was the 'central' technology of the industrial age (1934/1967:14). It will be shown that clock-entimed machine technologies became a dominating element of the social timescape and the dominant temporal rhythm of modernity, displacing the earlier forms of temporal relations that individuals, societies and cultures experienced prior to the clock's ascendancy.

It was in the context of this developing modern timescape, in the first Empire of Speed of the 18th to the late 20th century, that the political processes of liberal democracy were born. It was a form of politics that reflected the domination of the clock in the social, economic and intellectual timescapes that were its foundational context. Nevertheless this was an essentially 'conservative' form of political thought that did not have the 'in-built' dynamics of speed and change that characterize capitalism. And so by the 1970s liberal democracy was incapable of synchronizing with the emerging hyperspeed of social and economic acceleration that is at the centre of the capitalist network society. We will see that the second Empire of Speed that neoliberalism and the revolution in ICTs has made possible, functions far too fast for the institutions of liberal democracy to functional in any way that could properly be described as democratic.

This is a perception that has had many different political responses. A major response (or 'pathology'), as we shall see, has been a growing political apathy or quietism, where people seek refuge in consumption, and in the illusions of individualism. Others, however, continue to think in terms of the historic obligations of liberal democracy and search for political agency in new ways. Importantly these are ways computers and networks to bring together ideas, forms of resistance and options for change that are indicative of people 'desynchronizing' from the fast, instrumental pace of the global economy. Using ICTs in ways that are not oriented towards economic efficiency or as commodities for consumption is in itself a political act today. The political act of blogging, for example, can be a different form of politics, a part of a larger dialectical movement against neoliberal globalization and its colonization of the times and spaces of democracy. The shape of this politics is unclear and its future unknown. The aim of these politics are likewise diffuse and it effects are hard to gauge. However, in many ways they are they only game in town because neoliberalism and the free market stands alone, with no plausible alternatives on the horizon. And so we need to understand the very deep role that temporality plays in the creation of our Empires of Speed and this might enable us to think about how time and temporality can be reclaimed by people and cultures as social creations and not as a force that need be linked indissolubly with profit and acceleration and an anti-democratic capitalism.

* *A note on the irony of 'efficiency' and 'flexibility'*

In this book I use the terms efficiency and flexibility and productivity a great deal, as they seems to me to be expressive of key political, economic and technological forces in our networked, neoliberal and globalized world—and often in practice achieve the opposite of their intent (hence the irony). Rather than place each and every entry in ironic quotation marks, I ask the reader instead to take the un-quotationed use of these terms and read the ironic intent into them.

CHAPTER ONE

A NEW EMPIRE

Précis

Speed (as a function of time) fills our lives and helps shape the kind of world we live in. This has been the case since the dawn of civilization. But today a new form of speed permeates society: the speed of the computer networks that drive commerce and underpin the processes of globalization. It is an open-ended form of speed, which means that the rate at which humans communicate and the rates of increase in productivity and efficiency can never be fast enough. In this postmodern economy the rate at which we do things has become the defining factor. This is because under neoliberal globalization we live in a world where the commercial imperative of 'market forces' has seeped into every realm of life, binding speed and profit together as never before. The promise of speed, though, is a false one. As the world gets faster, then what Paul Virilio calls information 'gridlock' becomes an increasing problem. Under neoliberalism, however, the solution to problems such as this is yet more deregulation, yet more (and faster) information technologies—and so the spiral goes on.

In this book the term *Empire* now has another meaning, one that uniquely extends the traditional political dimensions that inhabited earlier forms of empire. The Empire of Speed is an empire based upon *technological systems* that are geared towards acceleration and neoliberal globalization. The present Empire of Speed based upon computer-driven acceleration is one where there is no one in control because politics can no longer synchronize (keep up) with the pace of change that has become an end in itself.

This first chapter acts as a series of signposts that touches lightly upon the theory of the Empire of Speed. It prefaces a set of arguments that will be developed in detail throughout the book. The chapter thus acts as a *mise en scène*, a highlighting of a logic which today envelops us but we take for granted—the need always to work faster, move faster, think faster and live faster. We mostly accept this 'social acceleration' and we mostly try to keep up with the pace of life. Rarely do we understand or question why this must be so. Even less do we appreciate the consequences. Once we have made the issues and problems of speed and time more salient, the rest of the book is devoted to making them *explicit*.

Move with the times…

Our planet Earth spins on its axis at a rate of one full turn in almost every twenty-four hours. The actual speed of revolution has probably changed little over millions of years. Ancient peoples would have had little idea of the precise physical dynamics of this processes, but the regularity, boundedness and invariability of this speed (limit) has been a source of immense comfort and great significance since our very earliest beginnings. We see this, for example, in the ancient astronomy of Mayan or Aztec civilizations of Mesoamerica, or in the stargazing British Druids who around 2500 BC constructed the megaliths at Stonehenge. More recently (and rather more influentially) our view of the world has been shaped by the scientific 'laws' derived from the movement of the planets by Isaac Newton in the seventeenth century—and who saw God's hand in the celestial 'harmony' he observed. Accordingly, entire cultures and totalizing worldviews have been erected on this basic physical movement in time and space.

Through the technological means they developed and through the political and economic contexts that help shape them, human cultures and societies have been moving (or at least have been perceived to be moving) at an *increasing* rate of speed since the dawning of civilization. This is a process that went into relative overdrive with the technologies and economic impetus of the Industrial Revolution (Adams, 1904; Simmel, 1907). Under the influence of industrialization, increasing social acceleration tended to stem from the capitalist techno-economic processes that acted in such as way as to pull culture and society, gradually, into its temporal rhythm, a rhythm that came to be regulated by the tick of the clock—the central technology of the age (Mumford, 1934). Much of this thinking was reflective and abstract and pursued by professional philosophers; the mass of ordinary people, by contrast, were largely born into this growing torrent of acceleration and for them it would have constituted a simple fact of life that was probably never much analyzed—except as a kind of stress, the provenance of which was unknown.

Today, however, we are *aware* of the shared experience of speed as never before. This is because the tempo of the pulsating dynamics of a globalizing world economy affects *more of us* at the *same time*, than ever before, and *more intimately* than ever before. Acceleration is almost palpable. And constant acceleration—the rate at which our experience of time has *broken free* from the temporality of the clock—is the defining process of our postmodern, post-Fordist, and post-industrial age. This rate of acceleration still issues from the techno-economic energy of capitalism, but it no longer *gradually* pulls the broad realms of culture and society into its orbit. Rather, the dynamic of velocity, acceleration, urgency, momentum or whichever noun we choose to employ, *suffuses* culture and society to an extraordinary extent. Acceleration stems not simply from new industrial processes, but also *rises up* from the ways in which we interact and communicate across all walks of life.

The postmodern 'acceleration of just about everything' as James Gleick (2000) has termed it, is no exaggeration or collective hallucination: it is our pressing reality. The rate at which we do things (or are expected to do things) both new and old, is now always oriented towards the fast end of an open-ended continuum of speed. We suffer now from a 'fetish' for speed that declares unambiguously and in all cases that the faster we do something, the better it is (Postman, 1993:19). Neoliberal economists and politicians call this speed obsession by another name. For them it is to be efficient. Moreover, the propensity for technologically enhanced efficiency is said to be what differentiates us from other species, is what makes us human. On a day-to-day level, to be efficient, as we are continually reminded by governments and employers and by the not-so-subliminal messages in the media, is to be on the cusp of the technological wave, to be alert to new opportunities, to be able to 'seize the moment' and squeeze it of its potential. To be efficient is also necessarily to be flexible—to be physically, cognitively, psychologically and metaphorically able to 'move fast' when the time comes. This may be a flexibility to change your job often, a flexibility regarding the way one does a job, a flexibility in your opinions (dogmatism is out for most of us), a flexibility with respect of your physical location, or even a flexibility in your physical appearance and how you construct your self-identity. To be efficient and flexible is to be able to move rapidly in response to 'outside' economic influences that constantly demand our attention. To be willing and able to move fast means that you can be 'successful' in your life, be able to 'synchronize' with fast-changing scenarios and rapidly unfolding events, staying 'ahead of the game' and hopefully out of trouble. The saying to 'move with the times' has never been more necessary or valorized than in our society.

Manuel Castells has called our present socio-economic structure a 'network society'—a globalized and interconnected society that is ordered and given organizational logic by computers. And machines, especially computer-driven machines, are efficiency, speed and flexibility *par excellence.* To substitute human agency with nonhuman computer technology is supposedly to decrease instances of inevitable human error (inefficiencies which cost money). The effect is that we are placed on a new and powerful system-logic that places vital emphasis on speed. The more speed the better, because fast is efficient. By *networking speed* we network and universalize this supposed efficiency. In the opening decade of the twenty-first century, we find the pursuit of purported efficiency through speed almost everywhere. To be outside the network is to be cut off from the spaces and times of economic opportunity (Lash, 2002). Accordingly, most of us consciously and willingly now choose to accelerate our lives, to seek greater efficiencies, devise more ways to be flexible because the alternative, so we are led to believe, is too terrible to contemplate.

Speed now dictates the forms and functions of the network society and networked economy. Prior to this revolutionary development (a time that was not so very long ago, it only seems like it), industrial economies were dominated by loosely coordinated Fordist systems of cultural and economic production (Harvey, 1989). Fordism functioned on the logic of economies of scale. The more standardized and mass-produced a product or service was, the more that could made for less cost, enabling prices to be kept low and the volume of sales high. This machine and factory based process had a specific tempo, and it was much slower than it is today. It took its time from the rhythms of modernity—and modernity and the Fordism it was to create as a mode of production, as a culture and as way of life, was synchronized by the clock. And like the clock, Fordist production and Fordist culture tended to be somewhat predictable. Choice was minimal because Fordism was relatively inflexible, and so innovation through competition took a relatively long time to evolve. The archetypal Model T Ford automobile, for example, its color as well as its technology, remained unchanged for fifteen years after its development in 1908. Such *stability* and long-term future in the life of a product—reflected as it was in the consumer culture that supported it—would be unthinkable today.

Since around the late-1970s we have begun to construct and inhabit a post-Fordist world which operates on what Rifkin calls an 'economy of speed' (2001:21). Time expressed as expressed through the economy of speed has supplanted the importance of space, expressed under Fordism as 'economies of scale'. Acceleration is the new prime mover. The ramping up of acceleration can be seen almost everywhere. Rifkin here observes one of its more banal features in the narrowing of the life cycle of automobile production:

> It took Chrysler fifty-four months, with a workforce of 3,100 people, to develop and manufacture its K-car in the late 1970s and early 1980s. A few years later, Chrysler developed its Neon automobile in less than thirty-three months, with a workforce of only 700. Today [2000] Chrysler... can deliver a new car in less than two years. Automakers envision that within less than a decade, they will be able to build and deliver custom-made, defect-free cars in only three days.

The increasing *rapidity* at which we produce, consume and distribute commodities is now the core process, the central factor in the 'economy of speed'. This represents an immense *intensification* of the pace of commercial, technological, and organizational innovation—and a transformation of the cultural and social forms that spin out from its epicenter (Harvey, 1989:147). In the economy of speed we are more than ever caught up in a race against time. The speed-imperatives of efficiency and flexibility function at the micro-level of the individual, and we all feel this keenly; but this 'iron law' operates also, and much more powerfully, at the level of the organization and the business enterprise. This is because 'being first to market allows companies to command higher prices and profit margins' (Rifkin, 2001:21). As Jack Welch, legendary business hero and one-time CEO of General Electric crisply put it: 'Speed is everything. It is the indispensable ingredient of competitiveness' (Lowe, 1998:47). The overriding objective for today's capitalist, then, is to occupy that narrow and evanescent time-horizon where the technological cutting-edge application or device brings the expectation of large short-term profits.

Getting to the market first with a new product or process was always the desired goal for the entrepreneur, but today it is the only goal, and you know that when you get there, someone will be right behind you. An example of this is Apple Computer's iPod MP3 player. At the time of its late-2001 launch it was the only product of its kind in the marketplace and Apple sold millions of them, making the iPod the biggest source of corporate revenue for what was then an ailing company. In early 2004 *Forbes.com* magazine reported that 'The would-be iPod killers are [now] starting to line up like ambitious fighters eager for a shot at the champ' (Hesseldahl, 2004). However, relative to the standard they had gotten used to, industry analysts saw such a competition-free time frame (around two years) as being extraordinarily long. When Sony's Network Walkman was finally released in mid-2004, a common market refrain was 'what took you so long?' (Chmielewski, 2004). The lesson was learned by Apple. In June 2007 the corporation launched its bid to become a force in the cell phone market with the iPhone. The device was hugely successful, selling over ten million units in a few months. However, so competitive and innovative (in terms of functionality) is the cell phone market, the iPhone had a shelf life of just over a year. With the July 2008 release of the iPhone 3G, millions of 1st release phones suddlenly became quite uncool.

What this demonstrates is the necessity to develop a high turnover rate in what is produced to compete in the economy of speed. In the age of globalization, the speed-imperative is, well, global. No company, no matter where they are or what they produce, is immune from the need to accelerate the rate at which they both innovate and produce. Hewlett Packard, a maker of PCs, printers and other high-tech instruments, which derives most of its profit from products that did not exist a year ago, further illustrates this logic. Beer is about as different as one can get from computers, yet the giant Miller Brewing Company makes 90 per cent of its profit from beers (brands and flavors) that no-one had ever tasted a couple of years ago (Rifkin, 2001:22). And the effect is even broader than the purely market span that separates beer from computers. From lager to laser printers and from automobiles to antiperspirants, the manufacture of almost anything today must be constantly speeded-up and changed. Nothing can be allowed to stand still, and individuals as well as corporations must relentlessly invent, initiate, transform, restructure, reorganize, revamp and renew.

A culture of speed

The idea that our cultural practices are being inflected and shaped by speed has fortunately been gaining more attention in the social sciences as well as in some general readership books. Sociologists Steven Bertman and Jon Tomlinson have both written excellent books on the subject. And looming as a presence in both these books is Paul Virilio, the French philosopher who has been speculating on the social and political function of speed since the 1970s. Virilio pervades Bertman's book as an uncited influence. However, the influence is more direct on Tomlinson, who praises Virilio's 'strikingly percipient book' (2007:60). I shall say more about Virilio later. What about the works of these sociologists on the connections between speed and culture? Bertman's 1999 book is titled *Hyperculture: The Human Cost of Speed* and Tomlinson's 2007 is called *The Culture of Speed: The Coming of Immediacy.* Both deal in great detail with the speed/culture nexus, and both are highly recommended. What they lack, however—and this an issue of focus as opposed to a intellectual critique of these excellent works—is a *theory of temporality* which explains *why* speed is becoming such a profound factor in culture and society. Both see capitalism and globalization as significant elements of the process of speed, and both take seriously the political dimension in that they see that political institutions have atrophied and that acceleration is running out of control. The problem with these books is that although they deal with the issue of speed they do not address the deeper-rooted question of time. They look at the effects of speed, whereas to understand the nature of cultural transformation in the Empire of Speed, we need to adopt a cause-and-effect that incorporates a systematic theory of social time. The subsequent chapters, I hope, will represent some initial moves in the building of such a theory.

What these writers do suggest, though, and I am in full agreement with them, is that the fetish for open-ended speed in culture and society has been turned into a norm. Put a slightly different way, the economic necessity of speed is held up as a virtue in our postmodern culture. And, indeed, the culture of speed is becoming dominant. Let us explore this almost intuitive notion a bit further. The culture of speed is a culture borne of the interactions of globalization, of neoliberalism and information technologies. And it is a culture that increasingly makes no distinction between work and leisure, private and public, day and night, physical and virtual. So deep has the process of internalization become, that often it is seen a badge of honor to speak of one's life as existing in the 24/7 society. This is a life where one's whole subjectivity blends into a flow of blurring and accelerating tasks. Obligations, incursions, commitments and projects are constantly juggled and foreshadowed towards a short-term horizon. In the 24/7 chronoscopic world that surrounds us, its signs and symbols, signifiers and referents, restlessly flicker and buzz to impress their urgency on our daily existence, compelling us to synchronize our lives to the increasing tempo of the overarching economy of speed.

From the media, from pressures at work, or at school, speed crowds in on us, reminding us of our inefficiencies and sluggishness and the need to get with it. The not-so-subtle implication is that if we don't learn new skills quickly then we will become obsolete or unloved or unemployable. Pop-up ads gatecrash our computer browser with the message 'your computer is not running at optimum speed!' and unsolicited emails will breathlessly inform that we can now 'fast track' a three-year university degree in two years or less—with term results sent to your cell phone. This is supposedly efficient, because accelerated learning synchronizes with the compressed temporality of the contemporary busy lifestyle, a lifestyle in which something like a home loan—which can spread out into the future for a quarter of a century and more—can now be approved and locked in during your lunch break.

The food we will eat during breakfast, lunch or dinner is increasingly 'fast': be it the instant oatmeal that is mixed with boiling water and ready to eat in seconds, the McDonald's hamburgers you consume on the move, or the frozen fish fingers you warm over in the microwave oven. In the economy of speed, rapidly produced, distributed and consumed food has become a major global industry since the 1970s. Pigs fatten faster, cattle beef-up in weeks rather than years, and battery hens lay more eggs when the lights are kept on for seventeen hours a day than they do if they live under more natural rhythms. Food would be far more expensive if the pace of production was not continually forced. Eric Schlosser sees our turning to fast food 'alternatives' as an 'Americanization of the world' (Schlosser, 2002). That it may be, but it is also an element of a general acceleration and monotonization of our food culture. A more certain consequence of this is that a fast-food diet has become the norm for increasing numbers of people. People resort to fast foods not because they taste any better than a wholesome meal prepared and cooked from scratch over a couple of hours: instead millions of people surrender their lunch hour or dinnertime in the quest for efficiency or flexibility. Fast food consumption thus functions as a coping tactic, a form of 'convenience' whereby time spent cooking can be freed up to deal with something more pressing, something more important than what seems to be the 'wasting' of time in leisurely restaurants or in the kitchen. 'Slow' food and broader lifestyle alternatives like 'downshifting', are for those few who can afford it (Tomlinson, 2007:10).

In almost every corner of the world the mundane commodities that give context and shape to our lives as consumers continually remind us of the need to do things more quickly. One only has to scan the shelves of the supermarket, that temple to consumerism, efficiency and supposed choice, to see the omnipresence of the message. Here speed is connoted with a sensuous desirability. Pick up your basket and walk down the aisles and you will quickly notice the bold tones of a *Rapid Shave* foam canister. This will be positioned to complement the *Mach III* blades (three times the speed of sound!) that will give the man a fast shave lest he misses that important appointment. *Speed Stick* deodorant will reside in close proximity, to give him 'maximum protection 24/7'. *Lady Speed* antiperspirant promises to keep women dryer for longer and *Slim-Fast* products will help both him and her lose those unwanted kilos swiftly—kilos possibly acquired from instant breakfasts, fast-food lunches and microwaved dinners. During the lengthening workday we can stay alert by consuming caffeine-filled energy drinks or by popping Provigil—'the first and only wake-promoting agent' (www.provigil.com). These may supply enough of the nervous and physical stimuli needed to get through a 'speed date' at a local bar where you can have 8 one-on-one 'dates' lasting 8 minutes each. And this life in the fast lane can be paid for promptly and efficiently with *fast cash* from the ATM or *instant credit* from your online banker. And if you run your own business, or if anything is tax-deductible in this frantic life, then *Quicken* software will make all these transactions go more smoothly (and quickly).

If still pressed for time, more technology is said to be the solution. Today communication through quick text messaging over a cell phone is massively popular because it is perceived to be faster and cheaper than writing or speaking. Broadband Internet is now touted as 'fast and reliable' whereas only a few years before, data transmission through the now-antiquated copper telephone cable was then also supposedly cutting edge and lightning fast. The US broadband behemoth, Qwest, publishes a business/culture magazine called *Lightspeed* that encourages users to 'ride the light'. What has changed? Just about everything—and the world keeps getting that bit faster every day. We shop and send mail and borrow money on-line because it's fast and easy and saves the time spent on a trip to the grocers, the post office and bank. Daily newspapers can be downloaded to our home computer or personal digital assistant (PDA). From those same network nodes we can work remotely when once it had to be in the office. Pointless journeys to the newsagent and office are now obviated and more time is saved. In our culture of speed, almost every product or service emanating from the economy of speed is now promoted as faster and more efficient with the corollary being that we will become faster and more efficient, more in tune and, somehow, lead better-quality lives.

The culture of speed has spread to the extent that today there are not many realms, or niches, secluded spots, or quiet spaces where the pace is not being stepped-up; fewer domains where what Thomas Hylland Eriksen (2000) calls 'slow time' is allowed to develop unassailed, and at its own pace. In the media, for example, adverts make a virtue out of the colonization of slower places. Clichéd images such as a person sitting on a lonely beach with a laptop that wirelessly connects you to the world is now supposed to be something appealing—as if one could happily commune with nature whilst reading a report or writing an essay, deleting spam mail, or trying to meet deadlines. But this is efficiency and therefore 'happiness' and therefore the whole point.

The developed economies lead the way in this colonization process and pull the less developed and 'slower' societies into their globalizing wake, transforming their cultures, politics and societies in the process. Sometimes the degree to which this logic is carried can be mildly amusing. For example in late 2003 the authorities in Ecuador launched what they called a 'punctuality drive' to make up for the perceived inefficiencies inherent that country's culture and economy. In a strong echo of Max Weber's analysis of the links between capitalism, religion and the time of the clock, Ecuador's business and civic elites chose to portray their fellow-citizens' relative ambivalence to the totalitarianism of the clock as a malady, a sickness that has to be cured, or a sin that has to be atoned for. Part of this battle for the hearts and minds of Ecuadorians was a nationally distributed poster campaign that read:

> Symptoms: Rarely meets obligations on time, wastes people's time, leaves things to the last minute, no respect for others.
> Treatment: Inject yourself each morning with a dose of responsibility, respect and discipline.
> Recommendation: Plan, organize activities and repair your watches.
> (Wilson, 2003)

The 'success' of promotions such as this has prompted governments other Latin American countries such as Peru, to take up similar schemes and exhort their citizens to synchronize with 'English time'. As a Peruvian cabinet Minister put it, the *Peru: La hora sin demora*, or 'Peru: on time' campaign is to get 'citizens in general, not just public officials, to undergo a change in mental attitude to start fulfilling their obligations' (Catan, 2007). Quite what these 'obligations' are was not stated, but clearly it is the obligation to synchronize with the economy of speed. For developing countries such as Ecuador and Peru, the writing is, if not on the wall, then certainly on the computer screen. To obtain that daily dose of temporal responsibility, to be able stick with the accelerating pace of life you now need (indeed are compelled) to digitally connect and interconnect. We must make ourselves part of the global economy of speed and make it a part of us.

The power of networks

Manuel Castells's heavy emphasis on the centrality of networks in *The Rise of the Network Society*, which was the first of his three-volume masterwork, reflected that fact that this was a key claim in his analysis of the emerging age of information. For Castells, networks are beginning to define society and, more, constitute 'the fabric of our lives' (1996:53). The increasing speed of networks, for Castells, simply increases their power to shape our lives even more, and in ways that we control correspondingly less (1996:32). Castells was writing in the mid-1990, prior to the generalizing of the Internet, the ubiquity of cell phones, the explosion of social networking sites (SNS) and widespread access to broadband computing. Looking back at his work today, it not only retains much of its analytical power, but also promotes 'networks' to an even more central position when we try to understand the nature of the changes ICTs have brought about.

Today, speed-generating devices such as cell phones, PDAs and laptop computers connect us to networks that annihilate space and compress time in a systemic and systematic way. Such technologies link to the Internet that acts as the backbone to connect hundreds of millions of users, organize production, develop processes, create services, produce and disseminate immense amounts of information and inexorably spread its digital domain throughout every realm of life. The speed fetish that governs computer research and development is daily exacerbating the problem, because R&D is oriented almost wholly towards making data processing and transmission as fast as possible. Research into non-commercial computer applications (so-called 'blue sky' research) is dwindling accordingly (Roush, 2007). Any new high-tech digital device must be immediately useful, networkable and fast if it is to be seen as relevant for consumers or business. From the latest MP3 player which connects to your computer to share music files with anyone similarly connected, to 3G cell phones that enable you to take pictures, watch TV, make short home movies and play and audio and video files, networkability is the key to whether the product or application will be commercially viable and successful.

The computer-driven engine of speed contributes to a world of speed in other, more profound, ways. The effect is salient in many areas of the hard sciences, which now also valorize acceleration. As we shall see later, the scientific control over time is concerned with the instrumentalization and rationalization of time's embedded diversity. This will be a recurring theme, but for now let me illustrate the argument with a short example. As just noted, science and technology increasingly seeks to *manipulate* temporality through *technique* and to insert speed into industrial and natural processes, and including research into the development of computing itself. So much so that the search for profit through speed has brought industrialization and nature together in ways (and through computing) that would have been unimaginable even twenty years ago. Whole new science and technology-based industries have emerged due to the quantum jumps in applied computing. This new digital power makes possible a bioengineering that is oriented towards speed and profit at the chemical/hormonal and genetic levels of life, and is a far cry from traditional selective breeding and intensive feeding in the food production industry. Greenpeace, the environmental NGO, reports that in the USA and China, experiments that mix human growth hormones with carp have shown 150 per cent growth increase in the size of the fish. And in China and South Korea, genetically engineered concoctions of mud loach and 'mouse promoter genes' have resulted in 'Increased growth and feed efficiency [and] 20- to 30-fold increase in growth' (Greenpeace, 2004). Usually such developments are routinely seen as positive indicators of efficiency and productivity and as science being put to the proper service of humanity. At a deeper temporal level, however, what this means in practice is that hormone and gene engineering cuts out the much longer-term process of the breeding operation. The waiting involved is not now across the generations of living creatures in the search to breed a larger fish, but the change is now instantaneous—and we have no way of knowing its future implications.

From the fundamental level of the hormone and genome to the broad dynamics of the global economy, the logic of speed of pervades our culture. Harvey underpins this idea with a political economy perspective in his *The Condition of Postmodernity* where he observes that 'the general effect…is for capitalist modernization to be very much about speed-up and acceleration in the pace of economic processes and, hence, in social life' (1989:230). Harvey was writing about what he saw as the nascent social and economic processes that were emerging from a moribund system of Fordism. He termed these processes a 'régime of flexible accumulation', an all-pervading force where once-discernible dualities of work-leisure, science-commerce, and economy and society begin to deeply intermingle. And flexible accumulation, as Harvey intimates, is much more than an economic process. It has become the basis of neoliberalism, the propellant for economic globalization, the spur for the revolution in ICTs, and is fast becoming the rationale for our being in the world.

The costs of no speed limit

Pervasive speed acts as a kind of opiate, a social economic and cultural addiction that it is difficult to break. Through the conjunction of neoliberalism and ICTs, acceleration now deeply enmeshes with our evolving social, economic and political structures. The important point to make here is that there are no objective limits to how fast we can continue to accelerate, and virtually no one in industry or in politics thinks about the consequences of this. We are driven by an open-ended continuum of speed where the point at which processes start to seriously break down are not known and hardly considered.

Paul Virilio has considered where this speed dependency will take us—indeed *is* taking us today—which is in many instances to a slow and grinding stop. In his book *Open Sky* Virilio formulates what he terms the *dromological law* that states that increase in speed is also an increase in the potential for gridlock (1997). Thus, for example, while automobiles, trains and aircraft have become faster and faster, the times spent in transit have not been compressed at a corresponding rate. Airports and train stations are points where increasingly fast technologies propel bodies through space. But they are also characterized by slow-moving queues and delays. Similarly with the Internet, that so-called 'information superhighway' where information, bits and bytes, images and data, travel at lightning-fast baud rates down fiber-optic cables or through the radio waves in the ether, gridlock is an everyday feature. Computers 'freeze', the network goes 'down' and 'traffic density' can maddeningly slow the rate at which you send and receive information. On a systemic level speed creates so much information in our network society that the result is what David Shenk (1997) calls 'data smog'—an information gridlock that can hold people in its grip just as effectively as being stuck in a traffic jam, or being at the mercy of a delayed connection at an airport or train station.

And, finally, at an acute empirical level, Virilio's dromological law was evident in the so-called 'credit crunch' that hit in 2008. Since the process of deregulation of the US banking system began in the early 1980s, individual banks were freed to pursue profits in increasingly innovative ways, through a broad process of what is termed 'financialisation', where the role of market-driven finance capital began to dominate all other forms of capital formation, i.e., industrial, manufacturing, R&D, etc. So intense has inter-bank competition (together with large financial institutions) become, that sort-term profit-seeking through increasingly complex lending 'products' were being offered to institutions and individuals with little or no concern regarding the ability of the lender to repay the loan. Since 2000, mortgages have been sold and resold and restructured in a growing whirligig of money-go-round, that as John Lanchester argues, has reached the point where no one any longer knows who is at risk and who is not (2008). This uncertainty eventually triggered an immense crisis of confidence across the whole financial system in the US in 2008 and this spread across the free-market world, which has also been deeply implicated in the process of seeking quick profits through speed with little concern for the medium- to long-term. Ex-Federal reserve chairman Alan Greenspan (and architect of much of this system whilst in office) unconsciously utilized Virilio's dromological law in a quote in February 2008, when, observing the general meltdown of the US financial system, said that 'As of right now U.S. economic growth is zero. We are at stall speed' (Reuters, 2008).

Temporal fluctuations such as these (go-stop-go on an individual level, and from fast-paced boom to sudden, grinding meltdown and back again) constitute the potential chaos and immanent unpredictability of our postmodern network society. The key consideration, however, is that *overall*, life and work and the speed at which the environment we create functions, *never slows down.* In a superb, aesthetically derived thesis on the nature of speed in our culture titled, *I Am a Bullet: Scenes from an Accelerating Culture*, Dean Kuipers observes that:

> Life will never move any slower. Nostalgia [looking back] leads us to believe that acceleration will plane out; achieving some static future state, but perhaps that state will be one of permanent technological and sensory flux. (...) Speeding up is what we do. It is human nature. We are tool users. We will always hunt for a way to do things faster, more efficiently, with less effort and less error. Does it ever gain us anything? Yes and no. Quantum leaps in quality of life come at a price: More tools we can't live without and which we probably have to work twice as hard to buy. (...) But constantly treading water at the surface of change has consequences. Deeper historical currents flow beneath us. We don't see them as clearly as we should. Sometimes we don't feel them at all...(2000:12)

Like a narcotic, increasing velocity never seems to fulfill its promise or live up to its glamour. Moreover as society gets faster, there is precious little objective evidence of increased personal freedom, of living in a world of efficiency and of contributing less effort for less error. This is because we are indeed 'treading water at the surface of change', or to use more expressive metaphor, we run like the hamster on the treadmill. We go as fast as we possibly can, and are pressured to go even faster, but we never actually seem to progress because we have no real and affirmative autonomy or agency over what are powerful, and networked *technological systems.* It is these networked systems that are the basis of the present Empire of Speed.

The 'deeper historical currents' that Kuipers speaks of are building today, deep and wide, into torrents. These flow, seemingly inexorably, across the whole of humanity and into the life of the individual, into the structures of meaning that comprise a culture, and into the institutions of economy and society. These 'historical currents' unite once-disparate realms, amalgamating them into an interconnected, fast-growing, fast-flowing postmodern Empire of Speed. In the chapters that follow, it will be shown that this is not an Empire in the classical sense such as existed in the Roman or British or Soviet archetypes. It is not characterized by the domination of one country of even one culture over an expanding territory of subject countries and cultures. It is a form of power and a transformed form of politics that has been characterized by Hardt and Negri as a '...*decentered and deterritorializing apparatus of rule* that progressively incorporates the entire global realm within its open, expanding frontiers' (emphasis added) (2000:xi).

Empire is a term indissolubly wrapped up in meanings of politics. It is used here to illustrate what the forces of neoliberal globalization and the revolution in ICTs have brought about. The present Empire of Speed is also a space-time filled with political processes, of course. But the politics are of a different, and rapidly changing, kind. Politics is and has always been a form of control and organization of human affairs, and since at least the time of the Industrial Revolution the political form with the most longevity and power has been liberal democracy. As I noted earlier, a principle argument I will make is that liberal democracy, over time, has developed institutions, processes, traditions and, importantly its *own temporality* that is based on the time of the clock. A thesis will therefore be developed which shows that network speed and time has begun to dissipate these political and cultural institutions of politics, organization and control. In moving from the temporal empire of the clock to that of acceleration and digital networks, we have undermined the liberal democratic institution and the cultures and societies that they engendered. Liberal democracy in its various forms in various countries can no longer keep pace with a fast-changing world and therefore the democratic process is less able to deliver on its historical responsibilities. That the 'loss of democracy' in its traditional forms is a *temporal* question is a relatively new concept and has only tentatively been dealt with in the social sciences (Reid, 1973; Virilio, 1997, 2008; Chesneaux, 2000; Rosa, 2003; Scheuerman, 2004). What will follow over the course of this book is an attempt to explore these concepts more fully by going fairly deeply into the nature of speed and time as a theoretical basis from which to argue the thesis of 'temporal empires' and the forms of politics that they engender.

In finishing this opening section, however, it may be useful to briefly sketch some of the more obvious linkages between temporality and political process, especially as they pertain to questions of individual and collective control.

Political control and network time

The potential for loss of control is an inherent feature in this generalized speed-up (Virilio, 1997). The neoliberal Empire of Speed is created by an increasingly ungoverned capitalism that has become uncoupled from the philosophical constrictions of modernity, from the technical restrictions of Fordism and from the local-political traditions of national sovereignty. Temporally unrestricted speed and spatially unrestricted capital are the digital Empire's motive forces. Hardt and Negri spoke of an 'apparatus of rule' in this respect. It is a suggestive phrase that begs the question: 'what is it?' and 'who wields it?'

When the cold war ended in 1989 and the rickety Soviet 'evil' Empire went the eventual way of all empires, the USA was supposed to take on the mantle (willingly or otherwise) of global hegemon, the enlightened repository of an idea of benevolent empire, of 'soft power' freedom, of choice and free trade. The process of globalization had been underway for over a decade since the fall of the Berlin Wall, when neoliberalism began to fulfill its inherent logic and become a planet-wide economic, political and technological phenomenon. Responding from a position of weakness, the 'liberated' states of Eastern Europe turned to the winds blowing from the West and opened their economies to dynamic flows of both productive investment and asset stripping; Russia itself was transformed into a variation of a market economy, as were China and India. The world seemed to be transforming itself into something like a mirror image of American capitalist 'democracy'.

As the most powerful economic country the US would be expected to dominate this new world as hegemon or benign power, and in many ways it does (Zakaria, 2008). However in other important ways the neoliberal network society is a society on autopilot and headed towards an unknown destination. Increasingly free flows of capital has meant that much US investment went overseas to buy up business, exploit markets and so on. At the same time, however, hundreds of billions of dollars flowed *into* the US market in the form of investment from China, Japan, Europe and elsewhere, creating a growing US dependence on this vast river of in-bound capital. Moreover, at the beginning of the 21st century, US government policy was creating the largest budget deficit in that country's history—a deficit that ballooned hugely as a response to the banking failures of 2008. This shortfall has also been built up through hefty tax cuts (bringing the level of government revenue through tax to the lowest levels in 2004 since 1950) and by massive military spending (Beams, 2004). And so, far from being the powerful and stable locus of a new empire of capitalist liberal democracy, the US is increasingly at the mercy of volatile global economic forces that have no real center of gravity.

Nevertheless, the US remains unchallenged as a military power, but efforts to use this to promote the US national interest for empire building or for any other professed reason in the wake of 9/11, has foundered in Afghanistan and Iraq. US military power has shown it can defeat any conventional foe, but apart from the dwindling group of neoconservative ideologues emanating from the Republican Party, US politicians, US capital, and the US public has shown little enthusiasm for the responsibilities or the consequences of the use of such power (Packer, 2005). Instead of being the confident and preeminent authority in the postmodern age, the only country who could feasibly lead the world in a benign fashion, or even dominate in it an overtly or covertly exploitative way, the US has been in fact reduced to being a deeply insecure and paranoid country (especially after 9/11). The biggest defense budget by far in many ways merely reflects the relative lack of national sovereignty and national self-confidence that the US showed in the decades after World War II. The political effect of a paranoia that stems from economic globalization is that in the early 21st century the US no longer sees the wider world through a lens of amity and cooperation (even on a rhetorical level), but increasingly 'through a gun-sight' (Pieterse, 2004:119).

The diminishment of national sovereignty in the globalized context has often been seen to be something that affects only the weaker economies, those who cannot resist the will of the International Monetary Fund, the World Bank or the World Trade Organization. In fact, the sinews of national sovereignty have atrophied *in every country*, including the US, as the ephemeral structures of a decentered and deterritorialized empire emerged. These structures are comprised of new constellations of power expressed as a volatile 'political subject that effectively regulates…global exchanges, [and is now] the sovereign power that governs the world' (Hardt and Negri 2000:xi). The postmodern Empire of Speed thus displays a completely new type of political and economic authority, which although bearing a resemblance to previous empires, in fact represents a transformation in the nature and operation of power (Kalyvas, 2003:265).

Today the primary 'political subject' is a growing range of national and supranational actors that vies for power and agency (market, social, political, regulatory and economic). These are actors such as non-governmental organization (NGOs), national governments, United Nations agencies, social movements; think tanks and, of course, business corporations. Power functions differently here, it functions erratically and unpredictably. NGOs and other political actors compete, collaborate, clash and conflict in the unstable settings of perceived interest. Gains by political actors in one sector can be just as soon lost or dissipated because in the overall political economy of speed, power has no solid basis on which to consolidate as it did in the long phase of modernity. Fast-moving circuits of information, commodities, signs and symbols (material, digital and representational) are characterized by '...their radical contingency and precariousness... [and] by the unforseeability of the sequences of events—sequences that are *always more brief or more compact temporally and thus ever less controllable*' (emphasis added) (Hardt and Negri, 2000:60–61). It is the supreme paradox that in a globalized economy where computers supposedly coordinate and make more efficient human domain over our environment, in fact, as Anthony Giddens observes in his *Runaway World*—'no-one is in control' (1999:7).

The nexus between neoliberalism and the revolution in ICTs have set this Empire of Speed into motion. The process of networking has made the rule of the commodity logic more powerful and pervasive than at any time in history. It colonizes spaces where it had not entered before, and creates *new* spaces of commodification and accumulation where none previously existed. By incorporating everything into its growing sphere, the Empire of Speed orients the entire social world to its needs. Indeed it enters into the very consciousness of people, ruling over bodies and shaping worldviews, constituting a new an inherently unstable 'biopolitical' power that has transformed the Foucauldian 'disciplinary society' into a society where institutional controls are yielding to abstract market forces. Seen from the perspective of temporality, and the speed context created by neoliberal globalization and the networked society, there is nevertheless a growing antithesis to this pervasive powerlessness, and this is evidenced in the nascent articulation of new forms of politics and new forms of personal agency and collective power. This discussion will take up the latter part of this volume. A start may be made, however, by looking briefly at the connections between politics and temporality as they have developed since the time of the Industrial Revolution and the phase of modernity.

Political control and clock time

The digital Empire of Speed deeply undermines the pillars of modernity that had been built up since the 18th century. Unconstrained speed and market-driven acceleration corrodes the ideas and practices of Reason and progress that had created a world of nation-states; it has swept away a Fordized industrial system that was based on long-term planning and cooperation between workers, management and government; and it has thrown to the dustbin of history the postwar experiment with social democracy. On the face of it, the suggestion that acceleration and modernity are antithetical may seem somewhat contradictory. Activists, artists and theorists from Karl Marx to Filippo Marinetti have centrally associated speed and acceleration *with* modernity, and as being the motive force of its development. And to an extent this is correct. For Marx, speed and capitalism were the very essence of modernity, a revolutionary force where 'all that is solid melts into air' as he memorably phrased it in the *Communist Manifesto*. The poet and dramatist Marinetti glorified speed as the dynamo of modernity, the source of power and of force that would act as a form of 'hygiene' that would sweep away the decadence and corruption of the old (and slow) world.

The form of speed that these and many other writers experienced needs to be seen in context, however. Marx wrote in the second half of the 19th century and Marinetti in the early years of the 20th. During these few decades, industrialism with its revolutionary machines and systems of power and speed were rapidly forming into the organizational logics of modernity, (Berman, 1981). These authors would have experienced a living and pulsating modernity unfold as a radical transformation of what existed previously. For Marx the speed and power of industrialism and modernity was oppressive, yet charged with liberatory potential for the working classes. For Marinetti, it signaled the means with which to transform the world into a more aesthetic (and proto-fascist) world, a new order where smoke, fire, steel, violence and speed would rule. The speed they experienced through modernity was real, and the tempo of life for them and millions of others was accelerated palpably.

However, this was speed of a specific kind, the tempo of which was circumscribed by technological, spatial, temporal, political, social and economic boundaries. These boundaries were the crucial 'material forms and processes' that Debray saw to be the forms that gave shape and temporality to the ideas of an age (2007:5). The clock served as the organizational meter around which industrialism, the science and technologies that sprung from Enlightenment thought—and ultimately modernity itself—was formed. Almost all modern technologies are speed technologies in that they have the object of 'saving time' built into their design. Modern technologies of speed such as machines are thus 'time-loaded'. That is to say, they were designed and developed with the exigencies of clock-time and work time built into them (Hörning, 1999:293–308). And in an ongoing process of replication, the time of the clock was 'folded' into the production of new machines and new technologies by machines and people that had already been 'entimed'. Through the constant use, development and replication of new technologies in work and increasingly in leisure, individuals, economies and societies began to synchronize, loosely, with the 'time-loaded' function of the technologies themselves.

The practice of institutional politics, of liberal democracy, was no different in this respect. As a social process, the temporality of politics reflected in general the temporality of the period of early and high modernity—and this was based around the 'conditioning' effect of the clock (Reid, 1973:201). The ticking of seconds, through minutes, hours and days; the rhythms of the machines it entimed; the schedules of work time; the division of time into periods of rest; the timetabling of transportation systems; of the introduction of Greenwich Mean Time as the world standard, and so on, constituted the thoroughgoing temporality of an increasingly modern and sophisticated and synchronized society. As will be argued in more detail in Chapter Seven, this clock-based temporality worked more or less effectively because the political process was able to function within the boundaries of these predictable and relatively stable temporal rhythms. For most of the period of modernity clock time afforded enough time for liberal democracy to deliver upon its historical responsibilities—again, more or less. And so in what constituted the exercise of political control over economy, society *and to some extent time*, futures could be planned; the past could be explored and mined for its lessons; and the present could be seen as one of active social engagement with the world, where people and the processes and institutions they created were able to make a difference on the world in ways that were more or less planned and intentional.

This was because the time of the clock is predictable and manageable in way the time of acceleration and speed of the network society are not. The clock, and the speed of modernity, constituted a loosely connected Empire or domain, what might be called a 'federation of speeds'. The spread of modernity was never comprehensive. Its relative lack of close interconnecting networks functioning 'on all registers of the social order' meant that historically there were always spaces and times where the force and speed of modernity did not reach—or at least took some time to reach. In the early 20th century, new technologies, such as motorcars, or trains, could hurtle through the cities and countryside to the amazement of their onlookers. Those who gazed at the spectacle could then go about their normal business once the machine (and with it modernity) flashed past them. However, those who moved rapidly in cars or trains were intimately affected in that they experienced a form of time-space compression that placed them in a different realm from those they sped past. For a time at least, their lives had become psychically and physiologically accelerated, they could 'do' more in their day than could those who remained relatively fixed in space and had differing experience of time. Cumulatively, other communications technologies had the same effect, and for correspondingly more people. For instance, the telegraph and later the telephone rapidly compressed time and space for its users, as these technologies rippled through society (Stein, 2001:206). And so for example a person in London speaking by telephone to someone in Sydney in the 1920s had inculcated into him or her a new and profoundly real 'consciousness of the world' as a singular place (Robertson, 1992:8). As communications technologies such as these began to suffuse society more generally the compression of time and space and the speeding up of time became a palpable factor for the majority of people in industrial societies.

However, even under late-Fordist clock time, differing and sometimes quite separate temporal realms could still exist within the lives of people. Picking up the phone to speak with a relative or colleague a thousand miles away was still relatively uncommon for most people (cost being a primary consideration in what were fairly scarce networks). Work could be kept separate from leisure, family from business, public life from private. The pace of life undoubtedly accelerated throughout the phase of modernity but the effects of speed were comparatively piecemeal and 'cumulative' as Stein (2001:201) puts it, and took time to spread across society in general. The crucial point is that the speed technologies of modernity did not network tightly and rapidly like digital technologies do today. It is only with the end of the system of Fordism and the transition to the current phase of post-Fordist postmodernity, are the shackles on space, time and speed broken.

The origins and effects of the present Empire of Speed is a quantitatively and qualitatively new phenomenon. It may be seen as an interregnum between two phases of history: between what had become a matured and increasingly static and lifeless modernity—and what will evolve out of it. The shape of the next phase of global order out of its present *dis*-order will be determined not by technology, but by people making political choices. Millions of people today, those who are part of the 'global civil society movement', as the final chapter of this book will show, reject the nexus between neoliberalism and the ICT revolution and the chaotic Empire of Speed it creates. More precisely, many explicitly reject the ideology of *neoliberalism* and the instrumental efficiency it supposedly stands for, and instead *embrace* digital technologies as potential forms of resistance and subversion. At least on some level of consciousness many realize that the Empire of Speed corrodes and renders increasingly ineffectual the classical politics of liberal democracy that grew and evolved with clock-entimed modernity (Scheuerman, 2001). As speed gridlocks physical bodies, then so too does it gridlock politics—not only in the politics of national sovereignty, but also in the day-to-day processes of democracy that touch the lives of everyone. Increasing numbers of people see civil society crumbling in the wake of this retardation of democracy and are attempting instead to forge alternative forms of politics. A broad aim is the construction of a new 'networked civil society' (Hassan, 2004) that uses global systems of communication to bridge the gaps between the local and the global and to undermine the dislocations of neoliberalism.

We will now begin to develop a historical and economic account of how and why the first Empire of Speed evolved. Just as importantly, however, we must also analyze its social and political dynamics to gain an understanding of the nature of the individuals and collectivities that comprise the countercurrents that refuse to run like the hamster on the wheel.

CHAPTER TWO

THE FIRST EMPIRE OF SPEED: CLOCK TIME MODERNITY

The experience of time's passage is intimately familiar...yet trying to explain to someone, who does not already know what is meant by the passage of time is not simply difficult, but seemingly impossible.

(J.T. Fraser, 2003:17)

Précis

Here we will try to simplify J.T. Fraser's dilemma and then shed some light on why this undoubted cognitive difficulty occurs. We begin with a more in-depth look at time from the perspective of the philosophers Bergson and Husserl. They represent a counter-Enlightenment view on the ways in which we experience temporality, by focusing on factors such as experience, intuition, memory and duration. Moreover I will utilize the work of social theorist Barbara Adam to build upon these perspectives—in part to develop a framework for understanding the social dynamics of time and our experience of it, through her theory of 'timescapes'. This theory represents a novel and powerful way of seeing time in a social context. It makes explicit what Fraser calls the 'intimately familiar' nature of embedded time and gives *entrée* into a revivified social science engagement that brings the temporal perspective in from the margins in the analysis of everyday life.

Outside of the world of philosophy and social science, however, a real world of technology and economy and human action continues to develop. Here time is unambiguous and dominating, it is the time of the clock, which is the time of money of industry and of technology. It is held up as commonsense time—as time *per se*. It is the form of time that displaces and makes obscure other forms of time, banishing them as unproductive and irrational. Since at least the time of the Industrial Revolution, the clock has created its own temporal reality and its own temporal empire.

A theory of social time

Modern social theory has taught us that time fundamentally exists within the social field; it is not an overarching cosmic universe, as Newton would have it, one that we exist *within*. Time is social, in other words. And being social, time is therefore also diverse. It emerges from differing cultures and differing contexts. We create and shape it. We are keenly aware of it on one level, and are blithely ignore it at another. At the beginning of this chapter I will illustrate time's inherently social nature by utilizing the works of Bergson and Husserl, two major philosophers of time, and then arrange their insights to build upon the social theory-based work of Barbara Adam and her concept of timescapes. This synthesis will form a unique theoretical basis to constitute the analytical framework that will make sense of our changing social, political, cultural and economic relationships with time over the last two hundred and fifty years.

Time is in us. There is, moreover, as Helga Nowotny (1989) puts it 'a time in everything' (*eigenzeiten*). In our daily life we weave in and out of these diverse timescapes with barely a thought to the matter. With little effort we navigate through the inherent temporal diversity, knowing them at some level of consciousness as integral parts of the wider whole that constitutes time. Nowotny argues elsewhere that 'time "dwells" in us—through the biological rhythms to which we are subject, and because we are social beings who are born into a society with changing temporal structures and learn to live in its social time... it is we humans who make time' (1994:6). We make time and we save time, we waste and spend it, we own it, buy it and sell it. As a society we use time as a form of reward or punishment: we can be rewarded (or can reward ourselves) by having a holiday a 'break' from the time of work; or society can seek retribution by depriving an individual of liberty, a time 'tariff' where he or she is locked away for a specified period.

Social theory, in large part derived from the Enlightenment and its ideas on the immanence of rationality, give insights such as these into the human temporal condition. Social theory helps us understand the centrality we place on time, whether it be as a reward, or punishment or as an abstract 'thing' in the world of commerce and business that we can trade, save or spend.

However, to say that time is social and can be explained in this rigorously rationalist sense (as revealing its whole 'truth') is only one fairly limiting aspect of the dynamics of time in society. To say that 'time is social' does not get us very far in conceptualizing what it 'is', how we perceive it, and how to distinguish between its differing qualities. This calls for another, less empiricist and more nuanced view of what constitutes our experience of time, and this is supplied in the works of both Henri Bergson (1859–1941) and Edmund Husserl (1859–1938).

The *experience* of time is a key term to try to comprehend in these perspectives. What these philosopher's argue is that in a world of positivist-based science and technology, what we experience as subjective beings tends to get lost in the noise and rush of the objectively based world 'out there', which is the basis of industrialism and democracy and modernity. So it is with time. Bergson's philosophy was aligned to a countertendency to the Enlightenment that included (most notably) Sade, Schopenhauer and Nietzsche (Bronner, 1995). He was concerned, along with yet others such as Alfred North Whitehead, to create an alternative metaphysics to what he considered to be the overly mechanistic and scientistic view generated by strict rationality. Bergson's fields of inquiry, consequently, were concerned with those subtle and hard to pin down areas that escape rationality's forensic gaze such as *irrationality*, becoming, memory and intuition.

According to Bergson, the primary awareness of what constitutes time comes to us courtesy of the Enlightenment and its creation of the physical sciences. This view apprehends time mathematically, as a series of fixed states that can be sliced up, separated out and measured precisely. In contrast to this perception Bergson argued that time is not something that can be wholly understood by number, but is something lived (internal not external) and durational, 'la durée' as he termed it. For example, he observes in his *Time and Free Will* that *durée* 'forms both the past and the present states into an organic whole, as happens when we recall the notes of a tune, melting, so to speak, into one another' (1991:100). Time, for Bergson, was therefore a state of becoming and a process of living duration. Moreover, the *intuitions* of the 'indeterminate' states of time were central to Bergson's philosophy. In his *The Creative Mind: An Introduction to Metaphysics.* Bergson described intuition as 'thinking in duration', a process that reflected the flow of reality (1946:129). Conceptual thinking *and* intuition were for him a necessary combination to perceive and understand the flow of temporality. Time, then, is not something understandable through a simple rationality that taxonomises, calculates, divides, measures and dissects. Time is indivisible and time flows. This fluidity is accessible, and is intuited as a lived reality. The value of Bergson's work is that he sought to separate out the qualitative temporality of the lived *durée*, the temporality of experience, intuition, memory and consciousness from that of the quantitative temporality of a rationalized time based on science, measurement and regular rhythms. Such a perspective opens up a way of explaining another, more evanescent and non-rational aspect of time which most of us experience (e.g. the way in which our experience of time does not easily fit with the ticking of seconds, through minutes and hours, etc.) but have difficulty in reconciling with the mathematical ordering of duration that classical science takes as the measure of reality.

Husserl was an almost exact contemporary of Bergson, and indeed their works share certain features that serve to compliment each other and take the perspective on time to a higher synthesis. Husserl is the founder of phenomenology, a philosophy that takes intuitive experience of phenomena as its starting point and tries to extract the fundamental features of experiences and the essence of what we experience. Like Bergson's intuitive method, Husserl's phenomenology is also opposed to Enlightenment-derived concepts of objectivism and positivism. His philosophical system was thus a descriptive analysis of *subjective* processes that has been described as the *intuitive study of essence.*

Again, like Bergson, Husserl did not accept the idea that time could be perceived (measured and categorized) as a series successive nows. He suggested that the essence of time is derived from our subjective experience of time. In other words, the essence of time is how we perceive it, not as a universal and absolute process as Newton would have it. As Scott Lash interprets him, '...Husserl enjoins us to begin not with the thought or the "I think", but with the "I experience"' (2002:102). The subjective quality of time needs to be understood. Bruno Latour (1997:172) has termed it the 'feel of time'. In contrast to scientistic conventions, Husserl conceived of the present a 'living present' as a flowing present, a 'now' in which impressions and perceptions stretch the mode of being through memory and expectations (what he terms retentions and protentions). 'Immanent contents' as he explains in his *The Phenomenology of Internal Time Consciousness*:

> ...are what they are only in so far as during their 'actual' duration they refer ahead to something futural and back to something past...we have retentions of the preceding and pretensions of the coming phases of precisely this content (1964:110).

Like Bergson, Husserl utilizes the metaphor of music to make his point on the durational and flowing nature of temporal experience. Memory and expectation, the stretching of the lived present into the past and the future, are central to the process of listening to music or playing it. Each note has a musical quality that depends on the location of the note in a whole flowing sequence of notes, the lived present, where the music is perceived. The isolated note does not have the same quality. As part of a flow of notes, the previous note retains its presence without actually becoming present. Similarly, with speech, or text, the word or sentence comes alive only when the preceding words are retained in the present and the future words, the 'coming phrases' (the expectation of future words) make the speech or text comprehensible. The past and the future dimensions of time are thus always in some sense in the present in Husserl's subjective experience of time (see Elizabeth Grosz, 1998 for more on this perspective).

If one can accept that time can be experienced and the experience be subjectivized; and if one can accept that time has a 'feel' and a durational quality that is unique to the experiencer, then the idea of *eigenzeiten*, that time is in everything, begins to take on profound meanings, and opens up the world to radically new interpretations.

The list of important time theorists is by no means limited to Bergson and Husserl. But they are valuable in that they wrote directly to the subject of temporality, and explicitly counterposed it to prevailing orthodoxies. Marx as the political economist and Weber the sociologist approached the subject more obliquely. The psychologist Carl Jung and historian Jacques Le Goff and E.P. Thompson also used temporality as frameworks of analysis with which to deepen and broaden their fields of inquiry. A fuller discussion of a greater range of theorists may be accessed elsewhere in the literature (see Adam, 2004). For now, though, this brief opening up of the differing possibilities of perceiving experiencing and analyzing time, gives us *entrée* into discussion of the notion of *timescapes*, a mode of analysis that seeks to give the widest possible interpretation of social time. It is a theory that builds on the major works of social theory and philosophy and seeks to understand more clearly *what it is* that is speeding up (and why this is so) through the invention and application of increasingly powerful and sophisticated technologies.

Timescapes

In the tradition of Bergson and Husserl, Adam's concept of timescapes similarly aims to break down the distinction between rationality and irrationality, divisibility and indivisibility, flow and phase, space and time, and, importantly, between culture and nature. In her books *Timescapes of Modernity* (1998) and *Time* (2004) Adam argues that we need to think of temporality, of temporal relations (the experience of the *durée* and a subjective and social process) as 'clusters of temporal features' (2004:143). It is a theoretical perspective that reveals a temporal world that is immensely complex—almost uncomprehendingly so. Complementing Nowotny's theory of *eigenzeiten*, Adam's idea of timescapes may be seen as the intricate intersecting of the rhythms, beats, sequences, beginnings and ends, growth and decay, birth and death, night and day, seasonality, memory and so on that constitute the diversity of embedded temporalities that are part of everything: from the eons it takes for a rock to turn to sand, the birth and death of a civilization, the life span of the fruit fly, to the lifetimes or minutes that permeate a memory or dream.

We are unavoidably immersed and implicated in these constantly shifting timescapes by simply existing in the social and natural world. Let me illustrate this idea: think of sitting in a park whilst reading a book, looking up occasionally to watch children play football. It could be any imaginable context, really. However, innumerable timescapes are connecting, breaking and reconnecting—and we are barely conscious of it. Countless relationships with time are interacting simultaneously. To be conscious of the context, to be what Ida Sabelis (2002) terms 'time aware', we can understand that, for example, the temporal experiences of the playing children are different to ours. For them it could be a feeling of time moving fast, or slow. They could be bored or excited, tired or full of energy; they could be anticipating the end of the game, or thinking back to mistakes they had made, or remembering periods of good play they had contributed to the game, and these may all feed into how they experience time. Taking our gaze away from the game to look down at a book allows us to enter another timescape. We can immediately become immersed in the story, become part of it and the images, scenarios and issues it conjures up. We may stop reading to close our eyes to enter yet another timescape by listening to the sound of our breathing, or the rhythm and throb of distant traffic. In our heads we can go back in time to think of a conversation that was had this morning; or project forward to a meeting with a friend of colleague. Opening our eyes and looking upwards to the trees that surround us we can see ancient branches move slowly in the breeze, marking their time with the variable strength of the breeze; or we can look down and watch the short stalks of new grass vibrate at a much higher tempo in the same wind. What emerges from all this consciously connecting with the temporal ecology of which we are a part, are the Bergsonian times of experience, of memory, intuition and so on, that were discussed earlier. We are a central part of this temporal diversity and we bring to it (and take from it) our own context of being in time. In the form of the subjective person, the timescapes we connect with in our everyday life ripple out endlessly and increasingly complexly into the social, natural and objective world—and ripple back in again to touch us. Moreover, our context touches those of others to create a connection, another context, another ring in the outward and inward rippling of social time, or timescape.

Context is the 'now' or the 'present' or 'becoming'. It is the intersecting point of contact between the different aspects of time that touch our lives—or those timescapes that we ourselves bring to a context or situation to generate a uniquely experienced temporality. As Christopher Prendergast has put it: 'What we call "the present" is a dynamic cluster of temporal traces, of the past it has been and the future it is in the process of becoming' (2003:99). What we create and experience in 'the present' is, in effect, a timescape that is part of a socially constituted temporal whole, part of what is to be alive in a becoming and emergent social world.

Bruno Latour constructs an imagery that is redolent of this complex interplay of temporal scapes when he writes that:

> ...all times converge [in this] temporary knot: the drift of entropy or the irreversible thermal flow, wear and ageing, the exhaustion of initial redundancy, time which turns back on feedback rings or the quasi-stability of eddies, the conservative invariance of genetic nuclei, the permanence of a form, the erratic mutations of aleatory mutations, the implacable filtering out of all non-viable elements, the local flow upstream towards negentropic islands—refuse, recycling, memory, increase in complexities... what is an organism? A sheaf of times? (cited in Bingham & Thrift, 2000:290).

As just noted, mostly we are unaware of our conscious self as being the centre of a whole concatenation of intersecting 'clusters' of temporality that touch us and we touch in mutual interaction. Nonetheless, within this ever-emergent and ever-moving timescape, we lead an incredibly elaborate and diverse temporal existence. We perform tasks, such as what we do in our daily work, we read, daydream, think, exercise, eat, sleep, concentrate, get distracted, dress, undress, cook, watch television, worry, talk earnestly, explain, chat, write and so on, in a limitlessly diverse number of ways and in uncountably different contexts. We may hardly give these times a thought or time may press heavily. It is a constant state of being-in-the-world that consists of the always joining and uncoupling, through shifting contexts, the incommensurable times and spaces. Most fundamentally it is the context that matters, the socially created intersection of times where events 'are "folded" or "pleated" into existence' (Bingham & Thrift, 2000:290–1).

However, we cannot carry this complexity, in all its growing fullness, into practical life; we can't think about this stuff all the time, in other words. This is part of the reason, perhaps, for the human compulsion to order and control time, to rationalize it and make it manipulable and predictable so that our lives can appear more stable and meaningful. The value of the theory of timescapes however, is to make this complex temporal world more apparent, and therefore more understandable. To enable us to do this it is not necessary to constantly view the temporal whole—indeed this is probably an impossible conceptual task. What we can do, rather, is to use a pared back version of viewing the world through timescapes to become 'time aware' by developing a specifically temporal frame of mind. The effect of this would be to make the times that are implicit in our temporal world more explicit when analyzing the context or situation or problem in question. The temporal dimension can give an additional perspective and thus a more holistic reflection of the dynamics operating in the social and natural environment.

Dominating times

Civilization, technologization and more generally modernity, have had a kind of dulling effect upon what could be a highly trained temporal acuity. Dominant or dominating forms of time have historically suffused and hegemonized our modes of temporal reckoning. The overriding mode can become the default form of what time is, to the point that we accept it unthinkingly *as time* and proceed to synchronize with its tempo. This governing form of time would dwell atop a hierarchy of times. For example, the preeminent temporal rhythms for the 12th century English peasant farmer could be, say, the rhythms of town, village, family, and home—and the innumerable timescapes that these contexts would continue to generate and the individual could 'live through'. The seasons were obviously fairly dominant rhythms in the lives of people involved with agriculture. Peasant farmers for centuries lived with and responded to the broadly cyclical prevailing timescapes of nature. Its rhythms of night and day, of planting and harvesting, of frost, sun and rain also featured prominently in their time-reckoning practices. This hierarchy of temporal rhythms would govern to a large extent the social rhythm of not only the individual, but also group and class, influencing other rhythms and sequences in the social sphere. And by synchronizing with this meta-rhythm of the season through habit, or by simply being born into its dominance, its temporalities would have become so customary, so second-nature that farmers would hardly think about it unless the rains failed to arrive, or unseasonable weather damaged their crop.

More recognizably, perhaps, we may begin to understand the concept of dominant timescapes and subsidiary temporalities if we consider being in an airport waiting to board a plane. The airport is the overarching temporal context, a context that is generated by our own personal life timetable and by the specific rhythms of the airport. The airport however is a highly elaborate technological timespace, governed by an abstract clock time that dominates everything within its orbit. It has precise flight schedules that need to be followed and that we need to synchronize with. We have virtually no control over this dominant public/industrial timescape. If we are too slow in getting to the airport, the flight won't delay for us. Mostly, of course, we get there in plenty of time—three hours for a long-haul flight—and therein we enter the complex timescapes of airline travel.

In the airport the flight-time hegemon looms over all, but within this preeminent time many other timescapes that exist in the hierarchy may begin to make their presence felt on our consciousness as we wait 'suspended' in time for the flight departure. International airports thus begin to upset many of the temporal rhythms that dominate daily personal and commercial life. For example, a fear of flying, or the celebratory impulse mean that people can be seen drinking in airport bars at 9 o'clock in the morning. Travelers who have arrived for their flight too early or have had their flight delayed or cancelled wander around listless and bored. Latecomers inhabit a frantic temporal experience, where the clock is running down fast on their scheduled departure time. They rush past those who have time to kill—individuals who protractedly write letters, read books or newspapers, scribble in diaries, or work on their laptops or talk into cell phones. They rush past overtired children and harassed parents who begin to wish it were all over. In this 'suspended' time, those with time on their hands can think of things in ways that depart from the usual mode: rewinding back in your mind to the recent time of packing and organizing for the trip, 'Have I packed my favorite tie, those black shoes I like, did I get the correct visa stamped in my passport?' and so on. Similarly, older memories can be delved into more deeply than usual, provoking anticipations, new hopes, anger, irritation, joy and anxiety in a swirling multiplicity of temporal/emotional experience that emerges under the ordering time of the flight's departure.

Once we step out of the airport at the flight's destination, however, the domination of the airline schedule evaporates and we suddenly enter new contexts and new connections and possibly new hegemonizing timescapes. Business affairs can take over, for example, insisting we be somewhere at a specific clock time which we need to organize and prepare for. Alternatively, if on holiday, we may find ourselves at a taxi stand with our backpack at our feet and with time suddenly on our hands. New contexts and new rhythms of life can suddenly emerge in new environments and whole new temporal experiences are now possible, with the complex entanglements of timescapes becoming something of which we are now keenly aware. And of course a biological timescape in the form of our diurnal rhythms can drag us back to yet another temporal awareness with the onset of a heavy bout of jet lag.

Social time is thus a universal flow of becoming that is comprised of the countless contexts of time that are colored and contoured, given shape, diversity, variation and substance through their unique circumstance of creation. The fact that this is social time indicates that these timescapes do not exist in isolation. Your timescape is not my timescape, but connections of similarity exist in society's web of connections. Social time, being comprised of the social, means that we share in aspects of it—the dominating time of the flight departure is one obvious example. However, society's connections go much deeper and wider, and so do our temporal connections, the 'scapes' of time that we share.

Industrialism and modernity created for the first time in history the prospect of a globally shared (and dominating) timescape, a timescape that is governed by the clock. The régime of the clock reaches into every culture and influences, to a more or less degree, the actions of most people on the planet. Moreover, this abstract, technological form of time, like all other technologies, *came to be a technology* because of the deep need for temporal efficiency and control, for the speeding up of time. The consequence of this imperative has been the displacement and/or subsumption of other times in individuals, in society and in nature. Let us discuss, then, the temporality of the clock, the 'engine' of the first Empire of Speed.

The industrialization of time and speed

The mechanical analogue clock is a relatively simple device. Compared with today's technical standards, its mechanisms are positively antique and watchmakers have great difficulty improving on any part of it as a selling point in the still-lucrative high-end wristwatch market. 'Distinction' is usually achieved through branding than through any intrinsic technological superiority of this watch or that. Nevertheless, this humble technology was profoundly ingenious in its logic and super-successful in its transformation of the world. It initiated transformations that run so deep that we have internalized them as normal. So much so that, in effect, the pervasiveness of clock time made our more ancient time-reckoning practices 'disappear'. This process of naturalization and internalization is, arguably, a major source of the dilemma that J.T. Fraser expresses at the beginning of this chapter. It is a problem that environmental philosopher Arran Gare conveys in another way when he observes that 'the permeation and domination of life by abstract [clock] time has become so complete that it is difficult to realize just how extraordinary this is' (1996:104).

What does the clock do? Through its mathematical gearing system, its design-function is to 'measure' time in society. However like trying to measure the waves on the seas, or clouds in the skies, or memories in the consciousness, and to attempt to measure these as regular and rhythmic sequences, it measures what cannot be measured in such a way. It slices the diverse timescapes of cultures, of societies and of the natural environment into uniform, precise and predictable bits that we can supposedly count and measure and value like pennies in a coin cylinder. Indeed, as we shall see, the coin metaphor is apt, because time is deeply connected to money and *vice versa*.

Why has it been so transformative? Clock time is a technological abstract form (abstract in that it exists neither in us nor in the environment that surrounds us) that is based on a scientistic view of the world and the universe. Its strict analogue logic cuts through the weaved intersections of times that constitute our ongoing temporal becoming. Instead it grids social and natural timescapes with a mathematical logic that produces a synchronization of the world in accordance with a Pythagorean *kosmos* where all things may be conceived of as number (Gare, 1996:76; Whitrow, 1989:38). The time of the clock acts as a powerful tool to which the times of people and other processes are coordinated. It attempts to graft the mechanical onto the social through an abstract numbering. This systematic structuring and synchronization of the diversity of social times goes back to at least the beginnings of the industrial revolution. The process was given formal institutional status, however, through the accords agreed at the International Meridian Conference in 1882. Here the world was mathematically and officially divisioned into 'times zones' that centered on the Greenwich Meridian in London. What this meant was that through the convergence of rational logic and an arbitrary political geography, the lives of individuals, cultures and societies would henceforth exist 'in' clock time. We would increasingly learn to look 'outside' of ourselves and our temporal contexts, to be 'told' what the 'correct' time was, and respond accordingly (Elias, 1992).

The clock is a technology of speed that has only one speed, but it created, or helped give the force of organizational logic to, the interrelated projects of Enlightenment and modernity. The clock has only one speed, but it injected *acceleration* into industrialism and capitalism as intrinsic elements of their logic. So much so, that in relatively short order, the clock, industrialism, capitalism, modernity and Enlightenment thought, became bound up within an irreducible and irresistible *force majeure* that was to become a whole way of life for millions of people through many generations in time.

A short step back in time

The clock has not always been so all-pervasive. Indeed it has a prehistory that is far longer than the relatively brief two hundred and more years' connection it has with modernity and industrialism. To more fully understand the dynamic relationship between the clock, the modern era and the relentless pressure to speed up, it will be necessary to take a brief look at the technical, social and cultural development of the clock.

The earliest European time-measuring device that had a shared rationale with the modern mechanical clock dates from at least one thousand years ago. This was the time of the so-called Middle Ages, the time of Alfred the Great in England, Otto the First of Saxony and Norman France. It was then that candle-clocks were introduced, notably by Alfred, as a means of dividing his day into roughly equal periods of prayer, royal duties, study and rest. The candle could be transformed from a light giving technology into a time technology by placing a heavy nail into the candle at the position where the required interval would come. When the candle burned down to this position, the sound of the nail hitting the plate below signaled the end of a particular duration of time. Although he probably wouldn't have seen it in quite these terms, but King Alfred's candle clock was a *productivity tool* that made his days more efficient, and no doubt served to speed up his perception of reality to the point where he could 'squeeze' more into his day by using his time in a more organized and protorational way.

Of course the candle clock is a technology, but it is not a mechanical clock. Mechanical devices (or clockwork) were explained in detail for the first time in Europe in Giovanni Da Dondi's *Il Tractus Astarii* (1364). Dondi's treatise described a seven-dialed astronomical clock that was regulated by an escapement of the 'crown-wheel-and-verge type' (Davies, 1997:434). The utility of mechanical clocks was recognized by the more forward-thinking elements of medieval ecclesiastical authorities who saw the obvious flock-control benefits in having people synchronize more of their day to this more efficient and predictable time reckoning process (Whitrow, 1989:100–101).

Accordingly by the 12th to the 14th centuries people in medieval Europe slowly became inured to a temporal regularity that could envelop the whole village, town or district through the ringing of church bells. Acoustic time signals emitted by bells were initially used to indicate prayer times and to call people to worship. However, as industry and commerce became more prevalent and social life became more complex, then so too did the reasons for time signaling increase. As David Landes notes, 'Bells [also] sounded for the start of work, meal breaks, end of work, closing of gates, start of market, close of market, assembly emergencies, council meetings, end of drink service, time for street cleaning, curfew and so on…' (1983:72). Time accuracy was of course rudimentary and the rule-of-thumb reckoning derived from the temporality of sunrise, mid-day and sunset, still drove the rythmicity of the medieval days until well into the 14th century (Thrift, 1996:191).

The subsequent introduction of mechanical clocks thus constituted a major revolution in systematic time measurement. It was with the invention of the escapement mechanism that regular timing could be signaled with a degree of independent (and abstract) reliability. Constantin Fasolt noted the extent of:

> the sensation caused in the 1330s by the appearance of mechanical clocks that were able to strike the appropriate number of bells at the appropriate hour, and the various ways of dividing the day into equal hours that emerged at the time: the Italian hour, counting twenty-four hours from nightfall to nightfall; the 'Nuremberg hour', dividing day and night into anywhere from 8 to 16 hours depending on the season; and the 'half hour', dividing the twenty-four hour day into two equal halves consisting of twelve hours each and beginning at midnight and noon (2004:415).

The deepening complexity of medieval life through the rise of industry and commerce, combined with the increased reliability of timekeeping through mechanical clocks, produced what Thrift calls an 'ideology of everyday time practice' that contributed to the formation of a system of 'time discipline' (1996:193). This was also a revolution in what it meant to be human in the social world, and may be seen as a foreshadowing of the onset of modernity. This nexus created a whole new dimension of social life that had evolved out of the dominating orbit of the sacred and feudal that had existed in Europe for at least half a millennia; it was a secular and instrumental realm that would constitute, in the late-medieval period (circa 1550s), the important structural supports for the coming revolutionary waves of industrialism and capitalism.

The sacred and the secular temporal realms began here to diverge. What medievalist Jacques Le Goff (1980) called the 'time of merchants' began to take its independence and autonomy from the 'time of the church'. The growing merchant class (the nascent capitalist class) increasingly saw time as a valuable resource though which commerce and production could be organized and profit derived. Merchants were the social revolutionaries of the period and they were shaking the deep foundations of an ancient feudal system as well as changing the time perspectives of those living within it. With the rippling out of industry-based time discipline across societies and cultures in Europe, a new temporal worldview began to emerge. And as David Loy argues, the idea of time '...began to shift from an *other-worldly* to a *this-worldly* dimension, an approach which no longer looked up to the heavens but forward to what could appear on Earth in the future' (2000:218).

And on the Earth this new and emerging mode of being represented the first systematic speed-up of life on a general societal level. It made possible at a very early stage a temporal-mechanical way of regulating the day, a process analyzed with brilliant clarity by Max Weber in his *The Protestant Ethic and the Spirit of Capitalism* that was written in 1905 (1989). Here Weber showed the social effects of the powerful convergence of religious ideas (its spirit or ethos) with the form of economy that underscored the Industrial Revolution. This was a profound ordering of the time of the day and enabled the intensification of what could be achieved within it. In short it allowed people to do more things faster. Activities could be planned and synchronized on a scale not previously possible. Deadlines could be projected and resources efficiently marshaled to meet them; systematicity could be injected into processes; and society could be organized and its objectives rationalized towards a singular (and secular) end. Above all a 'division of labor' could be organized in space and time, which as Adam Smith famously observed in his 1776 *Wealth of Nations*, massively increased the speed of commodity production (1776/1965:30). Speed, time and society began to inhabit a shared technologized plane. Moreover, within this plane they were oriented towards the same objectives of rationality in place of irrationality, organization in place of chaos, predictability in place of uncertainty and efficiency in place of waste and sloth.

Time and money

With the beginnings of industrialization and modernity the experience of space and time changed dramatically. Technological development was the objective driver of this process, but the most important transformation occurred within the heads of people. Enlightenment ideology began to construct a new rationalist worldview, a confident, forward dynamic that strongly promoted science and technology as the basis from which human society would improve, progress, and enable the construction a future in which almost anything was possible. In respect of how individuals and cultures related to temporality, the clock began to supplant the eons-old experiential (and diverse) relationships that humans had with time. Clock time became the dominant timescape through which temporality was experienced. In particular, through the convergence of the clock, industrialization and modernity, time was transformed from a mode of subjective experience into an abstract value. More, the correct *use* of time came to represent a necessary moral element of character.

This transformation is expressed nowhere better than in the connections between time and money. It was through the money economy that people came to see that 'profit and the more exact use of time are linked together' (Thrift, 1996:172). As the new secular domain of time grew in scale and complexity then the importance of timely coordination and synchronization grew in tandem. Le Goff (1980:30) explains the separation from religious time and the radical connection that it exposed:

> The merchant's activity is based on assumptions of which time is the very foundation—storage in anticipation of famine, purchase for resale when the time is ripe, as determined by knowledge of economic conjunctures and the constants of the market in commodities and money—knowledge that implies the existence of an information network and the employment of couriers. Against the merchant's time the church sets up its own time, which is supposed to belong to God alone and which cannot be an object of lucre.

Le Goff insists that clock time is at the very heart of the motivation for capitalist enterprise. The 'need for speed' is closely tied to the basic need for the capitalist to derive profit. And it was the elevation of the time and money nexus to central importance in human affairs in proto-modern European societies that gave powerful impetus to technological innovation. Indeed, during the 16th and 17th centuries nascent 'merchant activity' was being conducted on ground already fertile with numerous 'imperceptible revolutions' going on all across Europe in social, cultural and technological development. As Frances and Joseph Gies (1995:2) argue, the European Renaissance, the period prior to modernity and industrialization, can be seen as a period of:

> ...gradual, imperceptible revolutions—in agriculture, in water and in wind power, in building construction, in textile manufacture, in communication, in metallurgy, in weaponry—taking place through incremental improvements, large or small, in tools, techniques and the organization of work.

In other words the drive towards the speed up of society, of the networking and coordinating of commercial activity, the intensification of working practices and the increased transformation of the physical environment was not a sudden explosion in European societies. It had been gathering an inexorable momentum for generations. As the process of innovation became more integrated and complex, then so too did rapid technological development begin to move to the centre-stage of history. History was being driven by dynamic forces (such as) as the merging ties between technological development and commerce that made a fully-fledged capitalist system a possibility. The economic and social basis for the logic of capitalism to take root was being laid down. And as the layers of what Le Goff calls the 'information network' of 'economic conjunctures' became more intricate, competitive, faster-flowing and all-encompassing of culture and society, so a powerful feedback loop began to develop: the need to realize profit drove the need for technological innovation further; this, in turn, increased the need for yet more profits to fund the necessary investment, profits that were realized only through the development of more efficient ways to save time and money in the production process—through ways that could primarily be achieved through investing more into technological innovation, and on it goes. Once the logic of the feedback loop was in place the cycle would generalize and could, in theory, continue endlessly.

The linkages between time and money permeated the western consciousness so deeply, that it transformed the ways in which we 'modern' people began to think and speak and name things to describe the perceived reality of the world. In broad terms Michel Foucault sees a nomenclatural process at work which functions as a vital way for people to imbue a form of 'order' into culture and society, and thus make a specific kind of sense of them (1970). More pertinently, linguists George Lakoff and Mark Johnson in their book *The Metaphors We Live By* show how the trope of time as money is interlaced throughout the deep grammar of the language we use to describe and make comprehensible our surroundings. Symbols of time, efficiency, speed and money are now unthinkingly enfolded into our everyday speech. The examples they list are:

> You're *wasting* my time.
> This gadget will *save* you hours.
> I don't *have* the time to *give* you.
> How do you *spend* your time these days?
> That flat tyre *cost* me an hour.
> I've invested a lot of time in her.
> I don't *have enough* time to *spare* for that.
> You're *running out* of time.
> You need to *budget* your time.
> *Put aside* some time for ping-pong.

> Is that *worth your* while?
> Do you have much time left?
>
> He's living on *borrowed* time.
> You don't use your time *profitably*.
> I *lost* a lot of time when I was sick.
> *Thank you for* your time. (Lakoff and Johnson, 1980:8–9)

To this list each can doubtlessly add their own, more contemporaneous and demotic examples of how the language of time-money suffuses everyday life. For example, originating in the USA but now widespread, the term 'you're history' is used as a form of abuse to signify that someone's time is spent and that they are irrelevant (have no value) to both the present and the future.

When referring to the connections between the time of the clock and money, we are referring of course to money as *capital*, as the essential element of the industrial economy. Money and what it represented changed with the development of the industrial way of life. With capitalism, money became a commodity imbued with a specific value. Time and space are transformed through the 'circulation' of capital through the social, economic and cultural realms. Moreover, the ever-increasing speed of this circulation is central to the viability of capitalism itself.

The Speed of Capital

With the development of capitalism the 'imperceptible revolutions' in technological development became an interrelated and mutually reinforcing dynamic that enveloped more and more of society. A growing acceleration became its constant temporal mode, and the need to constantly expand into geographic and social space its overriding purpose.

Marx and Engels recognized this propensity as early as 1848 when they wrote the *Communist Manifesto*. It is an often-quoted paragraph from the *Manifesto* but its citing here is useful, not only because it shows the acuteness and prescience of their observations regarding the temporal dynamics of capitalism, but also because it reads like a freshly-penned description of the dynamics of neoliberal globalization in our own time:

> Constant revolutionizing of production, uninterrupted disturbance of all social relations, everlasting uncertainty and agitation, distinguish the bourgeois epoch from all earlier times. All fixed, fast-frozen relation-

> ships, with their train of venerable ideas and opinions, are swept away; all new-formed ones become obsolete before they can ossify. All that is solid melts into air; all that is holy is profaned... [national industries] are dislodged by new industries... that no longer work up indigenous raw material, but raw material drawn from the remotest zones; industries whose products are consumed, not only at home but in every quarter of the globe. In place of old wants, satisfied by the productions of the country, we find new wants, requiring for their satisfaction the products of distant lands and climes. In place of the old local and national seclusion and self-sufficiency, we have intercourse in every direction, universal interdependence of nations.
>
> (Marx-Engels, 1975:38–39)

Later, in *Grundrisse* published in 1861, and in Volume One of *Capital* published in 1867, Marx analyzed the socio-technical relations that comprised the central determinants for acceleration. These are labor-time and exchange value. Marx believed that in the capitalist process of production, labor creates value in two forms: use value and exchange value. The former is the capacity for a commodity to satisfy human needs; and the latter (the surplus value) contains the property of being exchangeable for other commodities. The relationship between worker and employer is one of exploitation whereby the employer must expropriate as much exchange value as possible from the worker. Time thus become pivotal to this relationship as it measures the amount of labor invested in the production of a commodity. Time itself instantly comes a commodity. Time is an inherent property of both worker and employer. However, within the social relationship of capitalism, the worker is compelled to sell or to hire his or her time at a rate set by the capitalist market, a supply-and-demand mechanism that is oriented always toward driving the rate down in the quest for the production of ever-cheaper commodities.

There is an equally profound *spatial* element to this process and that is the *circulation* (or actual exchange) of commodities. Time measures and commodifies the circulation process, making space and time hinge elements that affect—and are affected by—the capitalist social relation. How does this work? Ideally, circulation, or exchange, should be immediate so that profit can be realized quickly. In *Grundrisse*, however, Marx spotted a contradiction whereby the need for immediate exchange of a commodity invariably runs up against the 'fallow' or 'dead' time of warehousing, storage, sitting on shelves, distribution and so on. Summarizing Marx's theory in the *Grundrisse*, David Harvey notes in his 1982 book *The Limits to Capital* that:

> There is, therefore, considerable pressure to accelerate the velocity of circulation of capital, because to do so is to increase the sum of values produced and rate of profit. The barriers to realization are minimized when the 'transition of capital from one phase to the next' occurs 'at the speed of thought' (*Grundrisse*, p. 631)...Since an accelerating rate of turnover of capital reduces the time during which opportunities pass by unseized, a reduction in turnover time releases resources for further circulation (p. 86).

The history of capitalism is in part a history of class conflict and exploitation. But is also a history of the effects of the 'Constant revolutionizing of the means [the technologies] of production' as the 1848 *Manifesto* put it. It is a history, therefore, of constantly seeking ways to speed up commodity production and eliminate the drag on their circulation. Competition means that any successful method to accelerate production and circulation will soon be emulated in the marketplace and so the speed of capital gets ratcheted up another notch. The narrow criterion of efficiency dominates. Machines need to go faster and turnover rates must shrink. Speed becomes the datum-point not only of time but also of space. Commodities must not only be produced quickly, but they must also be able to move rapidly through space. It becomes an unacceptable drag on the logic of circulation to have, for example, a pair of shoes made by a machine in, say, three minutes, and then to have it take three years to come to be exchanged (or 'realized' as profit). That the process ideally should take place 'at the speed of thought' forms the technological basis for what Anthony Giddens calls 'time-space distanciation' (1990). This process transforms the objective qualities of space and time, thus compelling us to change how we act in the world. In sum: as capitalism becomes increasingly sophisticated, increasingly competitive and increasingly widespread through on-going industrialization, the world automatically becomes faster and smaller.

Modernity: acceleration as a way of life

As modernity and industrialization became the dominant paradigms in western societies in the mid-to late-nineteenth century, the apperception of speed through the political economy framework made famous by Marx and Engels began to spill over into other intellectual realms. Other theorists began to view these dynamics not simply as evidence of technological and economic transformation, but also as affecting an existential struggle in the minds of individuals. As Marshal Berman wrote, the gales of modernity that were sweeping the world comprised a 'maelstrom' of 'contradiction, of ambiguity and anguish' (1981:15). Some of the contemporaneous 'pathologies' of speed will be analyzed at some length below, but around the turn of the 19th century, these were previously unknown phenomena that were being analyzed by the equally new and mysterious disciplines of psychiatry and psychology. For example, in 1869 the American psychiatrist George Miller Beard diagnosed a new disorder of 'nervous exhaustion' or neurasthenia. This was in essence a speed-driven disorder that, according to Elaine Showalter, 'was the price exacted by industrialized urban societies, competitive business and social environments' (1985:137). A somewhat more influential psychiatrist, Carl Gustav Jung, had his own reflections on the temporal aspects of industrialization on the psyche. In his travels to the peoples and cultures of the non-industrialized regions of North Africa in the 1920s, he observed that:

> The European is, to be sure, convinced that he is no longer what he was ages ago; but he does not know what he has since become. His watch tells him that since the 'Middle Ages' time and its synonym, progress have crept up on him and irrevocably taken something from him. With lightened baggage he continues his journey, with steadily increasing velocity, towards nebulous goals. He compensates for the loss of gravity and the corresponding *sentiment d'incompletude* by the illusions of his triumphs such as steamships, railways, aero planes, and rockets that rob him of his duration and transport him into another reality of speed and explosive accelerations (1980:268).

In the broader cultural sense, what Marx, Engels, Beard, Jung and very many others were describing, was the *evolution of modernity* and its fashioning of individuals, cultures and societies into a wholly new and radical way of life. The dynamic logics of speed, commodification and spatial expansion became part of the human condition—part of what it meant to be modern. For Marx and Engels the unifying and uniform time of modernity would inevitably equalize humanity and eradicate the 'irrationality' of previous modes of existence that were bound up with religion, mysticism or feudalism. For Jung, the opposite was the case. Modernity was an 'illusion' that obscured other ways to be human to the point where we can no longer know ourselves.

For all of them, the clock and the speed that it generates were essential elements in the dynamic of change. But how does this work? How does the one-speed temporality of the clock generate speed and more importantly, how does it produce *acceleration*? It was shown how this operates through the dynamic of capitalism, through the speed mania that is fuelled by competition and the 'time is money' nexus; but there is another dimension here that is important to include. Arguably, most technologies are 'speed generators' in that their design and use are calculated to make processes more efficient and thus faster. It would follow that if speed is built into technologies then so must also be time. For example, the 'time' of an axe is immeasurable and not reducible to one time. It 'takes' its time from the context of creation. However under capitalism the time of a machine is *measurable* and is based on the time and tempo of the clock.

Bruno Latour in his essay 'Morality and Technology' (2002:249) describes the interactions of time, humans and technology:

> What is folded in technical action? Time, space and the type of actants. The hammer that I find on my workbench is not contemporary with my action today; it keeps folded heterogeneous temporalities, one of which has the antiquity of the planet, because of the mineral from which it has been moulded, while another has the age of the oak which provided the handle, while still another has the age of the ten years since it came out of the German factory which produced it for the market. When I grab my handle, I insert my gesture in a 'garland of time'.

To the temporal ingredients that Latour argues are 'folded' into technologies, there is another to be added: the *history* of the tools' 'becoming', the past that gives the tool its present reality. Think of a classic industrial machine such as a lathe used for turning metal. As Latour's hammer contains the age of the minerals from which the head has been moulded, and the wood which forms the shaft, so too, the components of the lathe contain the effects and traces, the history of the convergence of social, political, ideological, scientific and technological contexts that brought it into being. The lathe is a precision instrument whose rationale is constructed through the same mathematical rationale (of measurability) that conceives of all things as number—the same logic that divides the flow and emergence of time into seconds and minutes. It contains the effects and traces of a designer instilled with the heritage of Enlightenment thinking, a rational cast of mind and a planned and narrowly-conceived instrumental perspective, where what the machine will do in the future is knowable and predictable. The machine has a planned 'lifetime' measured in years. Moreover, it contains too the logic of the industrialism of which it is a part, that is to say, of commodity production, of competition, of efficiency and of speed of operation.

The lathe, in other words, is not conceivable without the intersecting tangle of the differing contexts of modernity that made it thinkable and actual. The lathe is the latest in a *generational development* of lathes and related machines that have *evolved* down this instrumental and narrowly delineated line.

The time of the clock is thus 'folded' into the lathe, into all the technologies of modernity. Indeed, as Karl Hörning *et al.* (1999) argue, technologies are inescapably 'time loaded' with 'temporal norms built into technical artifacts that not only determine the technology itself but also affect the daily uses of the artifact' (p. 294). The authors go on to state that:

> ...time is part of our physical involvement with the natural and technological world; our practical daily involvement with the material world is temporal to its core. Time is then more than a socially constructed category that allows us to move through the world in a coordinated and meaningful fashion. Instead humans establish particular temporal relations with the world; they are locked into the temporal and material world they create (p. 294).

During the period of early, high and late modernity, clock time was the dominant 'temporal relation' in the industrializing world; it was the controlling factor that created what Hörning *et al.* (p. 294) call the 'modern patterns of time' where 'humans are locked into [its] rhythm'. The use of the lathe and any other modern machine, unfolds its 'folded' traces into social, economic and political effects that become part of the generalized logic of modernity that emerge through Le Goff's 'networks' of 'economic conjuncture'. It is a logic that unfolds into the ongoing construction of the world according to the principles that gave them artifactuality. The clock is thus the motor and the tempo modernity. It 'time loads' technologies that are conceived, built and operationalized on a mathematical, rational and instrumental basis.

Acceleration *rises up* from this regular and equable basis. Acceleration is constrained only by current technological capacity, technological levels of development and, importantly, the prevailing economic and political tangencies. Acceleration is an open-ended *effect*, but the clock remains the overall (and underlying) coordinating, organizing, rationalizing and scheduling principle. The clock generates the necessary uniformity, predictability and planning capacity without which a dynamic economic and productive system based on pure and unbridled competition would quickly degenerate into chaos and collapse.

It was this clock-based temporality and the logics of measurability, predictably and mass-reproducibility that it engendered that became the classic markers of the productive force of Fordism that were outlined in the first chapter; indeed these were the essence of Fordism. This mode of production brought the logics of rationality and number-based concepts of Newtonian physics (clock time), science and technology, together with Enlightenment-based ideas of progress and increasing material benefits to its most developed form in what might be called 'high Fordism'. This was the postwar period from around 1950 until 1973. The first Empire of Speed reached its pinnacle during this period and it represented the zenith of the domination of the time of the clock in the lives of people, cultures and societies. Fordism, had become a dominant way of life, and the mechanistic rhythm of the clock underpinned its temporal patterns. Politics (and political institutions) have an especial role within this context and will be discussed in some detail in chapter six.

Fordism, however, had its intrinsic limitations insofar as capital's 'need for speed' went, and it is therefore a prime example of how economic and political tangencies can act as limiting factors (Debray, 2007). Fordism was at root a highly managerial and bureaucratic method of employing and deploying capital. To a significant degree, production and consumption were planned and scheduled and routinized under Fordism, with the clock as its base-line rhythm. It was by no means the operation of 'pure' market principals of free competition. It was politically managed to a large extent through tariffs, price controls and a state say in what a national economy produced and in what amounts. What this meant, in part, was that the speed of capital (the acceleration inherent in modernity) was kept in check through measured and prudent introduction (or non-introduction) of technological processes (such as computerization) that made production much faster, but tended to cost jobs. It also meant that the space of capital (geographic expansion) was limited in each developed country through instrumentalities such capital export controls, fixed exchange rates, and restriction on the amount of foreign capital that could enter the country to buy up local industries.

The 'natural' tendencies of capital accumulation are inherently spatial and temporal. Spatially, they are to expand as much as possible into new markets, develop new commodities and seek fresh sources of raw material and cheap labor to produce them; temporally, as we have seen, competition drives circuits of capital towards an open-ended acceleration, where faster is always better. Accordingly, Fordism, especially in its 'high' form that materialized during the postwar period in the Anglo-American economies in particular, where government, big business and organized labor comprised the bureaucratic and political triumvirate over the 'natural' operation of capital, came to be viewed more critically as forms of 'interference' and as unnecessary 'red tape'. This perspective became more prevalent as profit rates fell and levels of accumulated capital (with no suitable outlet) rose (Kolko, 1988). On the global scale this led to a pressure cooker situation in which the neoliberal demand to free capital from state and labor controls increasingly grew. Economic crises grew more untenable over the 1960s and 1970s as Fordism matured and ossified under the strains of its spatial and temporal constrictions. David Harvey (1989:187) writes of the spatial dimension to the growing crises, but the speed of capital is similarly deeply implicated in the process. He maintains that: 'the crises of Fordism':

> ...can be to some degree be in interpreted, therefore, as a running out of those options to handle the overaccumulation problem...As these Fordist production systems came to maturity, they became new...centres of overaccumulation. Spatial competition intensified between geographically distinct Fordist systems, with the most efficient regimes (such as the Japanese) and the lower cost-cost regimes (such as those found in the third world) were driving other centres into paroxysms of devaluation through deindustrialization. Spatial competition intensified, particularly after 1973, as the capacity to resolve the overaccumulation problem through geographical displacement ran out.

The 'resolution' to these spatial and temporal crises of Fordism was to be a full-fledged restructuring of the world economy on the basis of neoliberal ideas. This was the ideologically driven 'freeing' of capital from its constraints and so-called 'rigidities', to let it flow, expand, accumulate and accelerate as much as the 'free market' would allow. It is a process that continues today under the rubric of neoliberal globalization. In this contemporaneous form, the speed of capital and the space of capital had outgrown the ability of the political ideology of post war consensus-based social democracy to accommodate it. By unshackling capital from the technological, political, economic and cultural restrictions of Fordism, the 'speed of capital' potentially has no limits. Bill Gates, perhaps unwittingly, took up Marx's dictum of capital needing to circulate at the 'speed of thought' in his 1999 book *Business @ the Speed of Thought*. Gates comes to the book's main message early on when he writes that: 'Information technology enables you to act quickly' (p. 6). Here he was merely stating a post-Fordist neoliberal mantra from the 1970s which was that individuals, government, businesses, corporations, institutions needed to computerize (act quickly) or die. Across the world, corporations, small businesses and the public sector have all followed the neoliberal mantra with a revolutionary zeal. It is of course no coincidence that since 1975 Microsoft has grown from nothing to become one of the biggest corporations in the world, with the ICT industries more generally rivaling oil and dwarfing automobiles or construction in terms of economic importance. Computerization has transformed the clock-dominated world in less than a quarter of a decade, to the point where hardly a process is untouched by its influence and its speed. Fordism, as a dominant means of production and as a 'whole way of life', and the cultural, political and economic apex of the first Empire of Speed, is now almost extinct, and today dwindles in the twilight of history—a victim of the 'need for speed'.

CHAPTER THREE

THE SECOND EMPIRE OF SPEED: NETWORKED SOCIETY

Globalization cannot take shape without the speed of light.

Paul Virilio, 2000

Précis

The network society is the speed society *par excellence*. It takes capitalism up to a new level of technological sophistication and global domination. Digital capitalism represents a new form of production and consumption that is based upon flexibility and adaptability and what Castells sees as a new form of social organization 'aimed at the suppression of space and the annihilation of time' (1996:471). In other words, it is oriented towards pure speed. This chapter looks at the dynamics of this networked society. It considers what precisely drives it and what the economic and social effects are. Based upon the instrumental inflexibility of Leibnizian binary code, the network society has developed into a 'closed space', closed to all but those flexible and efficient processes and applications that are geared toward the production commodities and the seeking of profit. It is a network of networks that is given 'life' through the infrastructure of the Internet, and increasingly sophisticated forms of computing, which serves to colonize other realms of life and other ways of thinking. Importantly, this closed and colonized realm, this Empire, generates its own form of temporality, a 'network time' that is a qualitatively different form of time from its technological predecessor, the time of the clock.

Digital networks: the intensification of complexity

As an ossifying Fordism was being transformed and reenergized, the governing time of the clock was being supplanted. Today, neoliberal globalization and the information technology revolution, processes that detonated into life in the late 1970s as a solution to the perceived inflexibility of Fordism, are rapidly creating the second Empire of Speed. All empires inevitably crumble, or shrink, or become irrelevant and wither away. And so it is that the temporal empire that was metered by the clock, and helped generate the Enlightenment, modernity and industrialization is now being superseded. The rule of the clock as the even and constant regulator and scheduler of societies and industries around a world divided into time zones no longer fits with the new Information Age. The old model is now too slow in its maturity and has outlived its usefulness as a generator of profit. Today a new form of temporality is becoming dominant: this is 'network time'. It is founded on, and rises up from, the proliferation of digital networks. These are networks that are based on a singular logic, but importantly they also generate open-ended and multi-speed processes and applications.

In these early years of the 21st century, the pursuit of open-ended speed is unabated. It is a restless need for speed that touches everything and changes everything. To borrow from Marx and Engels who described the revolutionary vitality of the genesis of the previous clock-based Empire of Speed: ICT-driven speed processes '...must nestle everywhere, settle everywhere, establish connections everywhere' (1975:38). Marx and Engels had in mind, primarily, Western Europe and North America, whereas today 'everywhere' means just that—from the pristine temperature-and dust-controlled labs of Silicon Valley and Bangalore, to the rather more disordered and grimy temples of commerce in Shenzhen or London or New York or Jakarta, the dynamic of acceleration pervades—pushing individuals and societies forward and pushing them faster.

This transformation could not have occurred through the ideology of neoliberalism alone; neither could the world be as it is today, merely through the agency of innovation in the computer sciences. It had to be a joint social and technological revolution—one feeding the other and in turn being fed by it. The capitalist need for 'connections everywhere' to boost trade and increase production is given far greater scope by information technologies; the corollary being that the need to increase efficiencies in trade and production triggers massive investment into ICTs. ICTs today moreover are developed specifically with connectivity in mind—if it doesn't connect then it is essentially of no great use. The effect is that each application or device represents a connection to a connection to yet another in a process that ramifies across industries and through everyday life in the construction of a networked society.

Castells observed that: 'The information technology revolution, and the restructuring of capitalism, has induced a new form of society, the network society' (1997:xv). For Castells, the emergence of a networked society represents a Kuhnian 'paradigm shift' wherein ICTs take centre-stage as a flexible and invasive techno-logic that transform the range of older and more diverse processes and dynamics within the economy, society and culture into a singular connecting system (1996:61–62). This system is built on the need for capital to constantly expand. Through ICTs, digital connectivity allows computer-driven speed to enter the interstices where other times and other speeds once functioned, creating a totalizing system where nothing, in potential at least, is any longer outside its scope of operation (Eriksen, 2000:59).

It is the networked society that comprises the technological and ideological framework for the digital Empire of Speed. Again Castells provides the penetrating commentary when he writes that:

> Networks are open structures, able to expand without limits, integrating new nodes as long as they are able to communicate within the network, namely as long as they share the same communication codes (...) Networks are appropriate instruments for a capitalist economy based on innovation, globalization, and decentralized concentration; for work, workers, and firms based on flexibility, and adaptability; for a culture of endless deconstruction and reconstruction; for a polity geared towards the instant processing of new values and public moods; and for a social organization aimed at the suppression of space and the annihilation of time. (1996:470–471)

Digital networks incorporate individuals, institutions and organizations into a specific logic that takes them out of the abstract space and time of the clock. This new virtual space is a speed-filled realm, the temporality of which approaches that of a 'constant present', a networked timescape that becomes the governing experience of time. To understand this process more clearly, and to comprehend more fully what living in a networked society means, it may be useful to unpack some of the essential characteristics of network logic.

Digital networks

The currency of the network society is information and its language digital. The structure of this form of information expressed as an array of signs and symbols that stem from a pure mathematical abstraction, a binary logic that is the raw material, the underlying principle of the paradigm shift.

The computer is based on and functions through a precise method known as the binary system of numbers. A computer processes information, makes decisions and is able to proceed to the next stage on the basis of whether it reads the each stage of the code as being 'true' or 'false'. The informational core of the code is represented and communicated as strings of ones and zeros (e.g. 101011011). In a paper published in 1701 titled *Essay d'une nouvelle science des nombres* Gottfried Leibniz, a contemporary of Isaac Newton, outlined the science of binary code. His aim was to develop a way to transfer what he termed the 'laws of thought' that he saw to be subject to ambiguity and distortion when transferred through oral or written language, to an absolute and unequivocal mathematical form. He wrote that his binary language would eventually enable the *automation of reason* and become '…a sort of universal language or script, but infinitely different from all those projected hitherto; for the symbols and even the words in it would direct the reason; and errors, except those of fact, would be mere mistakes in calculation' (Bell, 1953:123).

Based on ancient Greek developments in arithmetic, mathematics and geometry, the binary system of numbers emerged as a method through which it was thought possible to apprehend the eternal 'truths' of the world. It is a rationale that stems from what Arran Gare sees as the predisposition in Western science and modernity for the 'quantification of human relationships'. As Gare argues, it was a deep-set ontology that was to culminate 'in the development of commercial capitalism…and an abstract form of thinking which denied objects their qualitative diversity' (1996:130). With numbers being at the very root of the Western worldview, this meant that the development of binary logic would have profound consequences for the diversity of meaning within human communication techniques. For example, through its modern application in computers, the pauses of indecidability in human communication (its 'qualitative diversity' to employ Gare's term) are increasingly flattened out and narrowed through their conversion to binary code. Digital transmission means that the latencies within the communication process—the times of pause and nuance and creation of a diversity of meaning through the variability of contexts—are being circumvented by speed and the yes-no logic of binary code. More generally the incommensurability and blurring of meaning that exist between languages, cultures, histories, times and spaces—all of which tended to militate against rapid information flows—are being homogenized and universalized by binary logic.

This pioneering work of Leibniz did not become a 'universal language of script' in his lifetime. Indeed it had been in a state of relative neglect, a byway in the larger discipline of mathematics that was traveled by only those few such as Charles Babbage and George Boole, who pursued the science of computation (Hyman, 1982). However, the increasing precision of 20th century machine and electronic technologies enabled Leibniz's theories of binary language to be employed in the construction of the first working digital computers in the 1940s (Edwards, 1995). The Second World War and then the Cold War that followed it had technological requirements in weapons development that gave tremendous impetus to the development of increasingly complex, powerful and faster computing machines. Accordingly, there followed a rapid technological evolutionary process that went into overdrive after the ending of Fordism in the 1970s. In this period we saw the rapid development of more sophisticated digital computers, the take-up and further development of these in the private sector, the development and spread of personal computers, the spread and commercialization of the Internet and, more generally, the eventual emergence of a highly interconnected network society (Castells, 1996; Schiller, 1999; Hassan, 2004).

It is the capacity to network that is the key function and the locus of social, economic and technological power in computing. We see this primarily in the capacity of the computer to dovetail with capitalism's need to produce commodities rapidly and to expand into space constantly. This energized fusion has been given much added momentum through a neoliberal ideology that seeks to tear down bureaucratic, institutional, cultural and legal barriers to the construction of a seamless global communication network within a free market context. It is a synergy that has created the foundation for the suffusion of a digital logic throughout society, a process that Mark Weiser and John Seely Brown term 'ubiquitous computing'. Through the growth of the Internet in particular, these authors see our own time as being '…an era of widespread distributed computing toward the relationship of ubiquitous computing, characterized by deeply embedding computation in the world' (1997:5). Ubiquitous computing thus an expression of a transformation of society where a single technology forms the basic organizing principle across cultures, industries, countries and peoples. And networking, as I argued in Chapter One, through Castells, is the central development of our time. To borrow from Castells once more, networks:

> ... constitute the new social morphology of our societies, and the diffusion of networking logic substantially modifies the operation and outcomes in processes of production, experience, power, and culture. While the networking form of social organization has existed in other times and spaces, the new information technology paradigm provides the material basis for its pervasive expansion throughout the entire social structure (1996:469).

The physical expression and social function of the technologies that comprise the network society give the outward appearance of variety and originality. For example, the 'ubiquitous' cell phone is available in many shapes and sizes and boasts a seemingly limitless array of functions and uses. Today, cell phones not only carry voice, but also send and receive text messages; they take pictures, play music, receive video, and give access to Internet and television content. They can also be used for playing games, as a diary, or function as a personal digital assistant (PDA) that can do all the standard things as well as send and receive emails, connect to your desktop computer or be used as a remote control for your television or DVD player.

The important point to be made, however, is that at their technological root, these (and other) ostensibly multi-functional and multi-applicational devices are the same. The restrictiveness of computer logic at its homologous core means that they cannot function beyond the mathematical inflexibility their binary code. Andrew Feenberg observes that as computing inexorably fills our lives its technologically derived 'narrowing' acts to 'distort' the 'structure of experience' (2004:86). In this 'narrowing' of experience, *actual* alternatives to binary logic are literally unable to connect to the network. This renders other ways of communicating and other ways of organizing outside the realm of the possible insofar as the design imperative of the network—and the neoliberal project itself—is concerned. Consequently the 'new social morphology' of our societies is not a digitally enhanced 'explosion of diversity' and innovation. Rather, the new digital paradigm represents a suffusion of the logic of rigid instrumentalism; a network of networks that does indeed 'direct reason' as Leibniz put it, but in a very specific way that has speed, efficiency, totalism and a sterility that blocks genuine innovation of a non-neoliberal kind, written indelibly into the code alongside its formal binary instructions.

Flexible networks

It may sound somewhat contradictory, but if digital networks are inflexible in their logic, they are also extremely supple in their own terms. This seeming inconsistency can be explained if the definitions of the word flexibility are explained and if we then look at what the social and economic effects of 'network flexibility' have been.

Flexibility is yet another classic economic platitude that has come to us from the 1970s. As Joyce Kolko has shown, it was a term employed to describe the basis of purported solutions to the problems of Fordism. It was seen as an antidote in that it was the opposite of Fordist economic and productive 'rigidity'—the gravest offence in the business context according to the neoliberal worldview (Kolko, 1988). Business leaders, politicians and economists who were sympathetic to neoliberal ideology maintained that flexibility had to be inserted into every business entity, every process, and every economic relationship. The call for flexibility is as strong today as it was more than a quarter of a century ago. This is because we are constantly reminded that there can never be enough of it, and that free market competition is necessarily a permanent state of revolution. The corollary to this is that being flexible is also a permanent state because competition in free market capitalism means that an innovation here means a necessary adaptation somewhere else—or sometimes throughout the system. To be 'rigid' and unable or unwilling to innovate ensures that you lose out.

The standard dictionary definition of the word flexible means 'responsive to change; to be adaptable'. The processes of flexible accumulation, therefore, are supposedly everything that Fordism was not, or was sorely lacking.

In the information society digital networks are hyper-flexible. Through digital networks, post-Fordist capitalism has developed and enhanced productive innovations such as just-in-time production, supply chain management, organizational flexibility, system integration, rapid-response time and the ability to operate markets at speeds approaching real-time (Goldman, 2004:2). Overall this these reflect a speed-oriented dynamic that is perfectly suited to the logic computerization. Moreover, flexible networks enable corporations to pursue the much sought after goal of 'weightlessness' where businesses are able to respond quickly to ever changing market circumstances. Here traditional physical infrastructure such as buildings, factories, production lines and large bureaucracies—which constitute such a drag on the processes of acceleration—are dispensed with through networked processes. For example, a British survey found that 'physical facilities will shrink by at least 25 percent in the coming years as firms make the transition to electronic commerce and to a network approach to organizational activity' (Rifkin, 2001:32). Many of the new industry pacesetters such as Amazon.com, eBay, Microsoft Corporation, Nike and Wal-Mart have become the prototypical models of weightlessness that others aspire to in the post-Fordist economy of speed. Of course physical infrastructures still need to exist to provide the actual commodities of consumption, but increasingly these processes are contracted out to producers in regions where the work is done in the most cost-effective way. Moreover, contracted companies are themselves tied into extremely rigorous production schedules where speed is of the essence. In this way, the dynamic of acceleration flows from the 'weightless' corporations in the USA and Western Europe to the fast and furious production lines of China, Malaysia or Brazil and Mexico, or to wherever 'outsourcing' is seen as cost-effective. The process of widespread computerization and the establishment of digital networks are central to this process. And as Weiser and Seely Brown imply their 'ubiquitous computing' thesis, almost anything in the design, production and distribution process—from clothes to art and from children's toys to buildings—can be computerized and subjected to the logic of networking and acceleration.

Flexibility thus develops as a flattening out of economic relationships that were once more diverse in time and space. It creates what Rifkin terms a 'commodification of relationships that extend open-endedly over time' (2001:97). Computerization is the technological handmaiden to flexibility that inserts speed into economic relationships in a way and with a force of power that has no historical comparison. It expresses the logic that can be traced at its deepest level to the binary language invented by Leibniz and developed further by computational theorists such as Von Neumann and Turing. As Kevin Kelly (1998) writes on the influence of the latter two:

> [these] mathematicians concluded that the essential process of computing was so elementary and powerful that it could be understood to happen in all kinds of systems...the notion of computation was so broadened so wide that almost any process or thing could be described in computational terms.

Prestigious institutions such MIT Media Lab are already putting the theory into practice by systematically searching for new realms of everyday life that can be computerized (Hassan, 2003). From retail sales to car manufacture and from education and medicine, the list of processes that have been enabled and made more flexible through digitalization is almost limitless. There is no need to spend too much time on illustration here; it is sufficient to cite only one, though highly instructive, example. Think of the humble analogue telephone that was the standard personal communication device as late as the 1980s. It was physically rigid—hanging fixed on the wall, or immovably on a table at home. When it rang you moved towards its permanent space to make the contact. If you were out of earshot, you'd miss the call; or if you got to the receiver too late, the caller may have already hung up. For all its usefulness, the old landline telephone was functionally rigid too. It dictated where you were located in space, and influenced your relationship with time. You were, for instance, reachable at a certain number only during office hours; and at other times only on the home number—if you were in. Out of the office and out of the home meant that a person could be almost impossible to reach.

Once made digital, however, the telephone is transformed and as users our relationship with time and space is revolutionized. *Mobile* phones (a much more apt name than the U.S. term cell phone) are now personal communication devices that unshackle users from the fixed point in space and clock time zones, making us contactable (networkable) almost anywhere at any time. They transform 'everyday' time and space by making them highly flexible; and they upset the more predictable rhythms of time and space generated by the fixed telephone by dissolving them into the virtual times and spaces of the 24-hour global network (Hassan, 2005).

The cell phone industry is one that thrives on the tropes of freedom, adaptability and the exhilaration of hi-speed networks—emphasizing in the process the more 'friendly' face of the ideology of flexibility. Adverts from Nokia, or Motorola or Vodafone invariably depict the world of mobile telephony as being the gateway to 'anywhere', 'anytime'; a world of (mainly) young people who represent a radical break from the dreary, predictable and fixed lives of their baby boomer parents—or those unlucky enough to be unable to afford the keys to digital heaven. In this virtual world, to be flexible is to be smart, cool and sexy. To be part of this world is to be someone who rides the cusp of the latest technological wave, someone who leaves the dumb, uncool and unsexy and unconnected far behind.

However the word flexible has more problematic definitions such as: 'susceptible to influence or persuasion; to be tractable'. And it is here, in the human realm, away from the fantasies of advertising and the alleged productivity gains and wealth creating dynamics of a globalized economy, where the real effect of flexibility makes itself felt. In the human economy of neoliberal globalization the suffusion of digitally enhanced processes and applications drives a fundamental transformation of social and economic relations that is characterized by flexibility of the kind just defined. In spite of the rhetoric, the networking and digitalizing of almost the entire world economy has not 'freed' people either temporally or spatially. Over the past quarter-century we have seen the emergence of a 'global elite' who drive and benefit from neoliberal globalization in the context of 'new speed—new polarization' (Bauman 1998:6–27). This elite use neoliberal globalization to enjoy 'unprecedented freedom from physical obstacles and unheard-of ability to move and act at a distance' (p. 18). However, the mass of humanity is *compelled* to conform to the flexibility and what Castells terms the 'fast-pace of the global economy' (2001:95). Indeed, many governments now legally oblige workers to comply with the new global reality through legislation in workplace relations of a kind that would have been unthinkable under the 'consensus' approach of Fordism. Accordingly, 'flexible employment' very soon became the sub-reality of 'flexible systems' as the neoliberal revolution got underway. And as Castells observes, this meant that:

> ...the notion of a predictable career pattern, working full-time in a firm or in the public sector, over a long period of time, and under precise, contractual definition of rights and obligations common to much of the workforces, is vanishing from business practice...(2001:95).

In its place has arisen a panoply of impermanent employment practices that exemplify what Nigel Thrift (2001:376) calls capitalism's need to 'continually create temporary actualizations'. Flexible hiring and firing policies, part-time work, short-term work, home-based work, self-employment, contracting, sub-contracting and sub-subcontracting now make up the complex and ever-shifting mix of working practices that are invariably tailored to suit the convenience of the employer. Networked capitalism is ideally suited to this hyper-flexible outlook. It allows for the continual experimentation of new ways of applying the logic of computing and networking; of new ways of speeding up processes; new ways to become more efficient and instrumental; and new ways to draft men and women into an increasingly exploitative and manipulative human economy that compels them to work longer, more productively, and more flexibly.

It is no coincidence that the ideal type of organization in these respects, the trailblazer for a whole new way of working, was the cluster of companies that make up the Silicon Valley economy. It was the nerve-centre not only of the new economy, but the fulcrum of the innovations in ICTs that made the new economy possible. At the level of the individual the general pattern to emerge and spread from the hi-tech, hi-speed, hi-profit ways of the Valley, was that the old-fashioned and quintessentially Fordist 'organizational man' gave way to the post-Fordist 'flexible woman' (Castells, 2001:95). Through the comprehensive domination of neoliberal practices, the theoretical idea of the perfectly tuned 'virtual' networked economy 'is able to take on flesh' in the form of 'flexibilized' men and women in the real world. Ultimately it is the mass of ordinary people who still must bear the costs of this flexibility. It is they that must respond to the ever-quickening tempo set by developments in computerization, and it is they who must chase the always-evanescent goal of happiness and harmony through the network valorization of speed and efficiency.

Colonizing networks

Related to the drive by digital networks to 'enable' and transform almost any process or application in the economy, and in culture and society more generally, is the underlying need for capital to constantly expand. The combination of neoliberal globalization and the ICT revolution has made it possible for capitalism colonize space and time with a rapidity and to an extent that was not possible under previous capitalist modes. Digital networks do what commerce and industry have always done, but they colonize much faster, they do it 'virtually' and with stealth, and they do it without the politically tricky physical incursion into the sovereign territory of a given nation state. Of primary importance is the fact that they colonize realms of life that were once outside the scope of capital in the search for new possibilities for commodification and profit. Colonizing networks are thus creating an empire that is unlike any that has existed before, and are suffusing the logic of commercialism so deep into the fabric of life that many do not see it for what it is: the colonization and pacification of space and time (and by extension, culture and society) in the interests of an abstract market ideal (Robins & Webster, 1999; Adam, 2004).

The extraordinary extent of the colonizing process has not gone completely unnoticed. For example, writing in 1996, William Robinson observed that it has involved the:

> …breaking up and commodifying [of] non-market spheres of human activity, namely public spheres managed by states, and private spheres linked to community and family units, local and household communities (1996:20).

Fredric Jameson made almost an identical point in an essay published in the same year where he argued that neoliberal globalization was inserting the dynamic of the market into social realms that had been '…hitherto sheltered from it and indeed for the most part hostile to and inconsistent with its logic' (1996:9).

Robinson, Jameson and other theorists see the process as unparalleled, but also fairly uncomplicated. They view it as a straightforward extension of the expansionary impulse of capital that has been in train since at least the 18th century. The analysis is achieved through the application of a more-or-less traditional Marxist political economy that views the world in largely unchanged macro-economic and political terms. Class oppression and class struggle is the standard cause-and-effect of the process, and 'solutions' to the intensive and extensive sprawl of globalization necessarily emerge predictably from this mode of analysis. However in such a conventional analysis there is no explanation as to precisely *why* the colonization process has gone so deep into the uncharted territories of culture and community, and why it has been so successful, so quickly, in creating these new realms for the introduction of market relations. The reluctance to factor in the effects of the crises of space and time in the capitalist economy, as well as the pivotal role of information technologies in its 'resolution', makes analyses such as these critically flawed. The far-reaching role of digital technologies is especially lacking. It is the failure to analyze this omnipresence, and the resultant flattening out of the traditional categories of class, that makes standard Marxist analyses such as these fall short.

It will be clear from my line of argumentation up to this point that the function and effect of ICTs is central. The success of colonizing digital networks lies in the fact that for most people such intrusion by the straitening and overwhelmingly commercialized logic of the cell phone or the personal computer, digital organizer or whatever, does not feel oppressive. It doesn't feel like part of a fiendish plan by shadowy corporate men in suits to make the 'masses' work harder for less. In what might be seen as the 'soft power' effect of global capitalism, empire building proceeds by dint of the active desire of the colonized. Nothing could be less oppressive-looking than a 'Blue Dalmatian' colored iMac computer, or the latest box of tricks from Nokia or Siemens. We have been educated to believe that to be connected to a hi-speed cell phone network or to work on a broadband or Ethernet link to the Internet is to be 'working smart', to be saving time. Representationally, in advertising and in media more generally, a powerful 'legitimation discourse' is constructed where '...speed is linked to the values of freedom...to the values of productivity, efficiency and control...' (Goldman, 2004:1). The flip side of this discourse is the idea that to *not* be connected is to fall into the dark chasm of the 'digital divide'. This is a place that no one wants to be in because it is not where the action is, and failure to be connected will see one's life-chances dwindle accordingly (Servon, 2002). Indeed, as UN Secretary General Kofi Annan argues, to be in want of digitally based information is now a breach of our human rights (Annan, 1999). And so we welcome the process of digitalization; we are schooled to demand it. The general message stemming from this is that to be 'disconnected' is a 'rights' issue, but to be connected is to move beyond that, or so it appears, into a realm of justice, a more equable and virtual level playing-field of opportunity.

The virtual time and space of the network has the *outward appearance* of flattening out class, region, city and nation state. And the beauty of this arrangement from the point of view of neoliberal ideology and the growing intensity and extensity of the network society is that we mostly subject ourselves to it willingly. By our voluntary actions we subscribe to the ongoing cycle of speed and connectivity. Whether it is an individual who feels pressed for time, or a corporation that is slipping in the competitiveness stakes, then for both, the now received wisdom is to have a faster computer, a cell phone, a dedicated link between office and home computer and so on. The corporation will invest millions of dollars in upgrading or replacing five-year-old IT systems that have become 'obsolete'. In the search for 'weightlessness' they will outsource their manufacturing, their administrative processes, or whatever it may be, to 'cost efficient' zones in developing countries thereby 'solving', for a time, their productivity and profitability problems. And in so doing they simply extend the network society still further.

Closed networks

Digital networks that are comprised of 'open structures, [and are] able to expand without limits' as Castells phrased it, contain also a very powerful *enclosing* logic. This seeming disjuncture can be explained. Digital networks do indeed extend and colonize every nook and cranny of the socio-economic corpus. However once the logic of colonization begins, its totalizing logic tends to close down those elements of economy and society that are not consistent with its modalities. If it does not close them down, then it marginalizes them and makes such practices, ideas, traditions, customs and cultures irrelevant to the main game of speed, connectivity and efficiency. In this the network logic mirror directly the operation of free-market economics more generally. For example, in the economic process, when tariffs are removed and capital is allowed to circulate freely, for to colonize and to expand, then inevitably small localized producers in retail, manufacturing, agriculture, and so on, cannot compete with 'globalized' and fully networked concerns. Over the last couple of decades, critical scholarship has shown the typical pattern of outcomes through such neoliberal openness. There seems to follow an inexorable pattern for those who are deemed outside the logic, which is: *incorporation* (being subject to takeover), *long-term decline*, or fairly *rapid extinction* (Teeple, 1995; Steger, 2001).

We have discussed the subtle, complex and ubiquitous nature of colonizing digital networks and the myriad devices and applications that constitute it. To most of us they appear as high-speed efficiency and productivity tools in our work and life. When we think about them at all, they are thought of as timesaving technologies as well as desirable cultural and design artifacts (commodities) that connect and enhance everyday life. This 'normalization' of the new has been a rapid and comprehensive process, with information technologies being at the heart of the developed countries' economic strategies, and a central developmental policy platform for the countries of the developing world (Castells, 1996; Foster & Seymour, 2000; Barney, 2004).

Just as the classical empires of old brought with them their singular worldview and justificatory ideology, neoliberalism and digital technologies combine to produce a powerful momentum that intrinsically precludes alternatives to it. Neoliberalism as an ideology and the binary language of ICTs are incompatible with other ways of organizing the economy and developing technological processes. Both are rationalizing and instrumentalizing forces that tend toward canceling out diversity and heterogeneity under the aegis of a globally networked economy and society. And like most empires, the process of colonization comes as Janus-faced. The inflexible face of the network society is revealed through reflective critique:

> We may see the development of the network society in terms of the extension of the [rationalizing] logic beyond the immediate labour process. Now it is the new information networks that impose the rules for socialization, dictating the logic of simultaneous collectivization and isolation. We live in an order in which all aspects of life—work, communication, leisure, consumption, education—are increasingly subordinated to the logic of the Information Society. The pursuit of rationality, efficiency and mastery has become a perverted form of rationality at the scale of the social totality (Robins & Webster, 1999:129).

This underlying reality of the capitalist digital enclosure, of the 'closed world' (Edwards, 1995) of network society discourses and processes, is masked by the kinder, gentler and much more attractive face of corporate and governmental ideology. This ideology—of a virtual world that is diverse and exciting—is inserted into the global consciousness and into individual worldviews, in large part by the arrangement of the global media industry that has a stake in such a worldview (McChesney, 2004). In particular, advertising by information technology corporations plays an important role in disseminating the global and uniform narratives that construct representations of a very particular kind of reality (Goldman, 2004). The universalized technical ideals of speed, efficiency, productivity and flexibility—intertwined with the social aspects of community, harmony, individuality and the 'cool' factor that utilizes sexiness and youthfulness—are the mainstays of a vast and perpetual marketing drive to draw the consumer/user/denizen into the network society.

Microsoft Corporation in many ways sets the ideological benchmark for the use of signs and symbols in the semiological battle over the meanings to be invested in free-markets and information technologies. 'Where do you want to go today?' was a slogan used by Microsoft to promote its software products on global television, billboard and Internet advertising. However, the freedom and choice implied in the catchphrase jars rather badly with the reality of the Microsoft domination of the PC software market and its colonizing strategies for the Internet. Microsoft, in other words, knows exactly where it wants you to go and where you most likely *will* go. The dominance of Microsoft and its alleged stifling of competition has been the subject of worldwide antitrust cases that highlighted the extent of their power in a globalized industry around which much commerce, production and culture now turn.

Another example is Nortel Networks, a global telecommunications corporation that uses similar strategies to infuse its products with specific meanings. During the late 1990s high-tech boom its advertising campaign around the slogan 'What do YOU want the Internet to be?' was complexly arranged to portray an exciting world of diversity and individuality that is immeasurably enhanced by being connected. Using the words of the Beatles song 'Come Together' their television advert portrays a connected planet of digitally-empowered individuals where it is they who construct the reality of the Internet, and will be the shapers of its future—not, it is implied, big and nasty multinationals. Also implicit in such ads is that whatever the Internet becomes it will be a mirror-reflection of their unique desires. (Goldman, 2004 et al.).

The cell phone industry represents a relatively new 'node' in the growing network of networks that comprise the backbone of the information society. It is a fully-global market dominated by less than a dozen corporations who between them sold 800 million handsets in 2005 to supply the surging replacement and new buyers market that make up the 1.7 billion users estimated in that year (Gohring, 2006). By the time you read this well over two billion users will have been connected—a figure representing almost a third of the total world population (Moore, 2006). Nevertheless, what Armand Mattelart calls the 'corporate ideology of global communication' (1995) is as relentlessly consistent here as it is in any other sphere of selling the ICT revolution. Speed, freedom, convenience, efficiency—with the added sublimations of sex and youth are the semiotic ingredients that work so well for companies such as Nokia, Vodafone, Ericsson and others in this immensely profitable and socially-transforming industry.

A clear pattern emerges from these brief examples. If the scope of meaning in the 'geostrategy' of the network creators seems narrow enough, then what they *in fact* signify also becomes clear if we critically analyze the process (Mattelart, 1991). The digital networks that individuals connect to are virtual spaces that they have no real control over. Corporate/commercial networks have devoured existing social spaces and created within them new spaces for profit that closes the user off to alternatives—or at least makes the alternative much less 'convenient'. Inexorably, then, the urge to text-message or email family, friends or associates begins to cut into the practice of speaking face-to-face to someone or writing down considered thoughts or reading a book. Before long, users in the network society consciously or unconsciously begin to synchronize their actions and thoughts to the accelerated tempo of instrumental reason. Being 'always on', in the technical parlance, means in practice that times of silence and aloneness with oneself and one's thoughts become an increasingly rare experience. Even a solitary walk in the countryside with a cell phone is still to be connected and available. The world and its concerns and its intensified information flows may still be pressed on you—and you on it. The 'always on' networked computer at home and work blurs the distinction between these spaces and the roles that are enacted within them. They help reorder these previously distinct and infinitely diverse roles in line with more instrumentally-geared *tasks* that are oriented at some level to the global system of production and investment that neoliberalism has created through the harnessing of the power of digital information technologies.

In a 2008 book by Jonathan Zittrain called *The Future of the Internet—And How to Stop It* the ubiquitous new networkable technologies of cyberspace are seen as a disaster for autonomy and innovation. Zittrain observes that early on in the ICT revolution, during the 1970s and 1980s, amateur tinkerers and computer enthusiasts could build upon the 'generative technologies' that were produced by the first-generation of network builders and Internet creators. Computer technologies of this period, stemming in part from the ethos of their creators, tended to encourage innovation, to be built upon and extended in different ways. Comparing the Apple II from 1977 with the iPhone of thirty years later, Zittrain writes tellingly of the loss of creative potential within the logic of commercial computing applications over that period:

> The Apple II was quintessentially *generative* technology. It was a platform. It invited people to tinker with it...The iPhone is the opposite. It is sterile. Rather than a platform that invites innovation, the iPhone comes preprogrammed...Whereas the world would innovate for the Apple II, only Apple would innovate for the iPhone (pp. 1–2).

Moreover, the encompassing and enclosing nature of digital networks is enhanced by its panoptic capabilities. The uniform binary language coupled with the colonizing propensity of network logic makes creates a closed virtual space wherein users are susceptible to constant digital surveillance by the owners and controllers of the network's protocols and proprietary systems. The closed space is patrolled and surveilled energetically by both private and government systems. And as Garfinkel (2001:93) argues: 'The big surprise...is that it isn't governments that manage the majority of this data, it's faceless corporations that trade your purchasing habits, identification numbers and other personal information just like any hot commodity'. Under neoliberalism, however, the 'big surprise' should be no surprise at all. Networked systems were designed from the outset with precisely these capabilities in mind: the creation of a virtual marketplace where every user is a potential consumer to be located, assessed and tracked in the virtual network—or, from the government perspective, is a potential threat to 'society' who needs to be monitored or physically found and detained.

Every user of networked technologies is thus a creator of data somewhere in the system; data that is stored somewhere and is retrievable by others. We leave a 'data trail' that is as easy for experts to follow as footprints in fresh snow. For example on the Internet there exists a software code called 'cookies'. This is a line of text that is placed on your computer hard drive which records and indicates on a file that you have visited a certain website. This file can be accessed by website operators to construct a profile of your web surfing habits, your general interests, and the sort of things you would be likely to spend money on. This data can be used by commercial entities as a way of targeting you for 'consumer surveillance' that leads to 'personalized marketing' by automatic systems (Lyon, 2002:43).

The network that purportedly acts as our gateway to freedom, efficiency and timesaving speed leaves myriad traces of our passing through it or acting on it every day—mostly in ways we are unaware of. A consequence of the network society's ubiquitous shadowing capabilities is that it functions very much as a 'surveillance society'. In potential at least, this digital Panopticon means that every keystroke we make on-line, every email we send or receive (however personal or embarrassing), every phone-call made or text message composed, every website visited, every word or phrase we put into the search engine, every download, upload and peer-to-peer file swap we make, every JPEG picture we send or receive—is vulnerable to relatively easy interception and capture. What you think, what you need or desire, how much you earn, spend, give away or save is capturable by computers. Data tracking systems can compile a dossier on your opinions, fears, hopes, illnesses and phobias. In fact elements of your innermost life that was once relatively bounded within the realm of your own individual existence and its small ripples of effect, are now available to others—anywhere and at any time—in the form of digital information. This can easily occur without your knowledge or permission (Lyon, 2002). And of course since the crumpling of the World Trade Center on September 11th 2001, not only big business has an acute interest in the intimate details of who you are and what you do—so also do state security agencies, armed with beefed-up legal powers of surveillance and souped-up computer systems (Rosen, 2005).

Somewhat ironically, the frenzied 'need for speed' in the network society, feeds directly into the increasing need for order, control and surveillance. In physics, increased speed can lead to instability and turbulence. It is the same with society. Accelerating speed, constant movement, perpetual change, short-term horizons, switching priorities, fiercer competition and the consequent drive to develop and introduce yet faster and more efficient technologies means that nothing can be taken for granted, nothing can be known for sure. The past blurs quickly into cloudy memories, and anything like a long-term future seems to be almost incomprehensible. Through speed, globalized and digitalized time and space are in a state of perpetual disorder. In comparison with today, the world of Marx where 'all that is solid melts into air' would seem positively rustic. Essayist John Berger, in his essay 'Against the Great Defeat of the World' sees the contemporary rampaging forces of technological and economic change as akin to painter Hieronymus Bosch's vision of Hell in his triptych of 1504, *The Garden of Earthly Delights*. Globalized society, he argues, is a 'space of Hell' wherein:

> There is no horizon.... There is no continuity between actions, there are no pauses, no paths, no pattern, no past and no future. There is only the clamor of the disparate, fragmentary present. Everywhere there are surprises and sensations, yet nowhere is there any outcome. Nothing flows through: everything interrupts. There is a kind of spatial delirium (1998:1).

The 'spatial delirium' that Berger describes, aptly corresponds to a profound disorientation in our temporal reckoning capacities. Jean Chesneaux claims that in our post-Fordist, postmodernity, the clock has been 'deconstructed', and that this negatively affects the functioning of the liberal democratic polity (2000:417). However, the effects spread much wider than that. The decentring of the clock as the lodestar of industrialism and modernity by the forces of neoliberal globalization and the revolution in ICTs has inaugurated a new and more complex relationship with temporality that is spreading throughout society courtesy of the ubiquity of digital networks. This new experience of time is called here 'network time' and it constitutes the emerging time(s) of the second Empire of Speed.

Network time

The global digital network is more than an infinite number of connections that enable people to communicate. It is more than a vector for binary information to flow through; it is more than a way for businesses to compress space and time in the constant search for efficiencies and speed; and it is more than an assemblage of technologies connected into a working whole. The network is alive. Literally. It lives and breathes and thinks and acts and reacts. It works furiously fast at a breakneck pace; or cruises at a more manageable speed, stopping to check, to rest, to reflect, to leap forward or step back. It is alive because the network is you, it is us; it is everyone who operates a connectable device or uses a connected application or process.

In his 2003 book intriguingly titled *Living Networks* Business consultant Ross Dawson argues that worldwide electronic networks constitute a '*global brain* in which ideas procreate freely and [within which] we collaborate to filter an ever-expanding universe of information' (p. 3). *Living Networks* is a book written for business people and for would-be NASDAQ investors and so somewhat predictably, much of the rest of book continues in this fanciful and uncritical fashion. As such it is typical of its genre, what Thomas Frank calls the burgeoning 'school of guru-interpretation' (2000:242). The idea of the network as a 'global brain', even as analogy, does not work as it suggests a centrality, a unity and an overall coherence that simply does not exist. Nevertheless, the notion that the network in some new way represents the living, technologized expression of hundreds of millions of people is useful and workable as a framework of analysis—as a way of viewing 'the body as information processor' and the network as the global extension of that body (Tofts, 1997:94).

Networks do have, to use one of Dawson's more acceptable phrases, a 'presence' (2003:115). This presence is virtual, a virtuality that is no less real for its being decentered and disembodied. What this virtual presence of over one billion people constitutes is a larger virtuality, a virtual environment or virtual *ecology*. The virtual ecology of the network is a 'real virtuality' as it involves real individuals who comprise part of it and works in tandem with digital systems on the reality of the concrete world to change and shape it. The virtual ecology is created, maintained and sustained as a consequence of the users' capacity to be 'always on', and through the system's logic that is oriented towards 'ubiquitous computing' which, in turn, creates the appropriate environmental conditions for what another business consultant, Jon Lebkowsky, usefully terms the 'persistent connection' (2004).

The network ecology is evolving out of the nascent forms that cybernetic interaction on a large scale are now making possible. As the next chapter will show, the idea (and the reality) of the cyborg (the fusion of humans with technological aids) has been with us for some time now (Wiener, 1947; Licklider, 1960; Haraway, 1991). Through theorists such as these, both science and social theory have made much headway in understanding the possibilities and implications of what a deepened, more entangled and completely intractable relationship we humans have with technology. However we are as yet far from the science-fiction stage where 'bits and atoms' combine (to use Nicholas Negroponte's phrase) into a proto-*Terminator* cyborg being that is living flesh covering powerful computing and hardened steel.

At the beginning of the 21st century we are in the preliminary stages of the evolving network society and are taking only the initial steps on what will doubtless be a long journey. Nevertheless, so intimate have we already become with cell phones, personal computers, PDAs and so forth, that it is already possible to view these as our cyborg extension into the network; technologies that unite us with it and it with us. And it is here that the socio-technological linkages are made which constitute the framework for an information ecology. The ecological framework is created and held in place by ever-densening webs of communication, the 'persistent connection' that creates a 'presence' or, more accurately a 'telepresence' that form the growing digital environment. No longer are the components of the network reducible to their component parts, i.e. computer, server, user, software, time and space. Instead these combine to form a living, amorphous system. They become the social and technological means of innumerable contexts (digital times and spaces) that individuals, groups, corporations, institutions, processes and applications blend into lived realities. Contexts may combine and separate in the space of nanoseconds—or last for hours or days or weeks. Contexts traverse geographic space, dissolving it into the virtuality that is the network society (Manovich, 2001:25–26). Importantly, this digital environment or ecology is as real as the built environment that comprises the cities and towns of our social world; and as actual as the natural ecology that provides the basic building blocks (the contexts) for life on earth. It exists because intense matrixes of connectivity create the framework for concrete actions that have real effects on people and on the interpenetrative social, political economic and technological systems they construct.

Using the Internet as his basis for analysis, John S. Quarterman (2002) argues that:

> The Internet is an ecosystem. It is composed of many interacting parts, ISPs, datacenters, enterprises, end-users, each of them drawing sustenance from the others and from raw materials. Each of them needs to make informed decisions. This is an ecology. And this ecology whose life forms are corporations and people is also a market.

Of course the Internet is one aspect of the network ecology. Our digital environment grows daily, hourly, and by the minute as people connect through the new devices and applications that come on to the marketplace every week, suffusing the 'persistent connection' ever more deeply into everyday networked life. Some theorists call the time of the network 'real-time'; indeed the term has been imbued into the popular imagination as standing for the speed of computer networks. The term 'real-time' is in fact a misnomer. Its connotations lend it to the construction of a general consciousness that tends to obscure the actual temporality of the network. A much more interesting and potential-laden time is 'network time'. This theoretical venture is tentative, but the preconditions for its fully-fledged emergence are already in place. As Jacques Attali (1982) put it:

> ...a new perception of time...is always preceded by the invention of the instruments required for its measurement and followed by the appearance of a new master of social organization...Conversely, the theory for each new time structure would seem to be never totally formulated until the end of the given structure's domination (cited in Klein, 2004:252).

With this in mind, acting through a human/machine-generated ecology of interconnectivity, our long-standing relationship with the time of the clock is not only being 'deconstructed' but is undergoing profound transformation. Hundreds of millions of individuals now inhabit—day in and day out, in their business and leisure, and in their public and private lives—a digital space where the time of the clock has less and less relevance. What matters now are speed, connectivity and flexibility. The times of the network are generated along a spectrum of compressed clock-time, a 'chronoscopic time' (Hassan, 2003). To be sure, the times that are generated in the new economy of speed tend to be fast, to be accelerated, but not on the plane of real-time. So before describing what this book sees to be the defining temporality of the network age it is still necessary to ask: what is real-time?

Computer programmers and systems designers coined the term to describe operating systems that could respond at high-speed to the input of data. The computer technicians' online dictionary of Internet terms defines real-time as something 'occurring immediately', and on a surface level at least this is how most people would conceive of real-time. However, this generalized definition, stemming as it does from a technical perspective, sheds little light on the social, cultural and temporal implications that 'occurring immediately' may signify. Michael Heim, in his *The Metaphysics of Virtual Reality*, gives a rather more intriguing definition. He writes that real-time is 'Simultaneity in the occurrence and the registering of an event, sometimes called synchronous processing…' (1993:157). This represents a significant shift from the technical definition. 'Immediately' connotes a brief temporal lag (be it measured in minutes, seconds, or even nanoseconds). Heim's 'simultaneity', however, suggests 'happening at the same time', a canceling-out of temporal duration, delay or latency between events. Simultaneity implies, then, a non-time, the shattering, or voiding, or 'death' of time.

A problem here is that social theorists and the media more generally, have taken the technician's term of indicating something that happens in digitally compressed clock time (fast, but still multi-durational, multi-patterned, etc.) and implicitly or explicitly take it to mean *no time*. For example, Castells, in his 1996 book *The Information Age: The Rise of the Network Society* argues that globalization and the information age are heralding the era of domination by real-time, or what he calls 'timeless time.' Real-time, for Castells is therefore also a kind of 'non-time' which means that as the network society becomes more encompassing of culture and society, 'linear, measurable, predictable time is being shattered...in a movement of extraordinary historical significance' (p. 433). In his more speculative social philosophy, Paul Virilio is even more explicit when he writes in that 'the teletechnologies of real-time...are killing "present" time by isolating it from its here and now, in favour of a commutative elsewhere that no longer has anything to do with our "concrete presence" in the world...' (1997:10).

If we think about the nature of time, however, we can understand that the concepts of 'timeless time' or of the 'killing' of time make no sense at all. Ontologically it is impossibility. We are temporal beings living in a temporal environment—whether inside or outside the network. As I have argued previously, diverse temporal durations, patternings and rythmicities suffuse everything: from the rapid heartbeat of a baby in the womb, to the several years it takes for pack ice to form. Like trying to imagine 'time before time began', i.e., before the Big Bang fifteen billion years ago, we evolved anthropologically and culturally ill-equipped to think in such terms.

We may more readily appreciate the absurdity of simultaneous real-time if we think about our own involvement with the network society. Think of the Internet. Its technical capacities and our own human capabilities ensure that this is an inherently *asynchronous* space. Nothing occurs instantaneously, or in real-time. There exists instead an open-ended temporal continuum within the network, measured from picoseconds upwards. For example, we can flash an email across the world in seconds or minutes, and then wait for an unknowable period for a reply. This could come in seconds, minutes, hours, days, or never. Networks can fail, they can slow down or speed up; we could be using state-of-the-art technology, or an old 486 PC and a dialup modem. The multiform temporal dimensions that we are able to create, at least in potential, in the Internet, has led Lee and Liebenau to insist that '...we can regard the experience of using the Internet as one of pseudo-instantaneous access' (2000:51).

Accordingly real-time and its social theory meanings of 'instantaneity' or 'no time' needs to be brought into proper technological and ontological perspective. Real-time may be viewed, therefore, as the final goal of machine/human interaction, the very end of the temporal continuum that stretches from 'no time' to the speed of light. To be able to achieve true real-time response would mean the ultimate surrender of human agency to digital technology, where latencies would have been driven out and where lags no longer occurred. This would constitute the militarist dream of the achievement of absolute power through absolute speed (Virilio, 1986), and the capitalist Nirvana where production and circulation function 'at the speed of thought'. Both dreams however are destined to be unrealizable as imperfect humans constantly get in the way of perfect systems.

Network Time as connected asynchronicity

What might be termed a 'connected asynchronicity' is a central feature of the network society and network time. These temporalities fundamentally change our relationship with the clock—they do not negate or cancel it. Instead, the numberless asynchronous spaces of the network society, created and inhabited by people and ICTs in interaction, *undermine* and *displace* the time of the clock. What we experience, albeit in very nascent form, is the recapture of the forms of temporality that were themselves displaced by the clock. That is to say, the embedded times in humans and nature and in society that formed the timescapes that intersected our lives, the timescapes we increasingly failed to fully experience, appreciate or understand due to the deadening implacability of clock time.

Asynchronicity connotes a temporal fragmentation, a smashing of the uniform and universal linearity of the clock into a billion different time contexts within the network. In a similar vein, Olivier Klein maintains that due to the effects of high speed and computer technology a network-generated 'fragmented time' is emerging alongside the 'industrial time' of the clock. Indeed, argues Klein, this 'new time structure, not totally formulated but already distinct' may signal a '...return to the "task oriented" work that Edward Thompson (1967) attributed to pre-industrial societies' (2004:252&255). Thompson's research indicated that within pre-industrial societies, before domination by the clock occurred, 'task oriented' time, i.e., the time it took someone to sow a particular-sized field or shoe a horse was dictated by the task itself and the interaction between human, technology and local circumstances. To time the work as 'one hour', 'sixteen hours' or whatever, would have seemed an illogical way to look at the job. The straightforward 'task orientation', Thompson concluded, was, to the pre-industrial worker, 'more humanly comprehensible than timed labour' (1993/1967:358).

Thompson's 'task oriented' time, Klein's 'fragmented time', and what I have termed 'network time' have important overlapping similarities. They denote a time that is created and shaped by a context that emerges through the intersection of humans, socio-economic forces and technology and not by an abstract and standardizing clock time. In the terms of the latter two theorizations, what is suggested is the creation of a new space for new experiences with times that are able to go beyond the purely instrumental 'task' that network designers initially set for them.

Within these digital spaces of asynchronicity we have the potential to create and experience our own context-dependent times, where the clock will have marginal or no effect. Think of inhabiting an Internet chat room, or conducting a three-way VoIP videophone call. It may be midnight where you are and mid-afternoon for your one of interlocutors; breakfast time for another. The conversation, however, takes place in network time. This may be fast if the network is running fast, faster if you have a whiz-bang computer and broadband fiber-optic access—or slow if network conditions are busy and you are using a copper-wire telephone connection and 28000 bps modem. The important point is that this *context-created temporal experience is disconnected from the local clock times of the users*. The clock no longer governs as it once did in the pre-Information Age. What this means is that the much more fragmented and contextual times to vie with 'industrial time' for our predominant experience of time for the duration of our presence online in the chat room, or on VoIP, or surfing of the Internet, or speaking on a cell phone, or whatever it may be.

Temporality, language and imagination

What we do when we enter the network, then, is to change the 'normal' governing dynamics of modernist space and time. By so doing we change the way the mind and body is potentially able to experience time and space. We have no sense organs for perceiving time, but our total awareness is able to become attuned to a greater temporal inventiveness. Such an effort at the level of our imagination is needed if we are able to take account in our dealings with a whole range of time-focused environments. We need to think more 'temporally' if we are to more fully appreciate (and exert some sort of control over) the times we create and move through almost unthinkingly.

Language already captures some of the diversity of time experience, and a more conscious appreciation of how we speak about time is a necessary starting-point to develop a deeper sensitivity to the temporal scapes that are part of being in the world. Consider the words we use to connote temporality and then think about how these conjure in our mind the different times we are able to experience. Imagine the forms of temporal potential within such terms as 'latency' and 'immanence' with its suggestions of not quite controlling or connecting; the physicality and textuality of 'pace' and 'intensity'; the uncertainty and indecidability of 'contingency' and 'context'; the drama, speed and force of annihilating time and space through 'distantiation'; the durational qualities of 'rhythmicity' and 'change', 'timing' and 'tempo'; the evanescence of words such as 'transience'; or the ethereality of 'transcendence'; the impossibility suggested by 'irreversibility'; and the complex and overarching interrelation between the influences of the 'past', the 'present' and projection into an open 'future'.

Gaston Bachelard in his *The Poetics of Space* noted this ability to change our nature and hence our awareness of space and time through the experience of environmental change. Principally Bachelard had in mind our relationship with space, but the logic equally applies to the temporal dynamics that tightly interpenetrate the spatial. He writes that:

> By changing space, by leaving the space of one's usual sensibilities, one enters into a communication with a space that is psychically innovating. For we do not change place, we change our nature (1996:15).

Quantum computing: the merging of bits with atoms

There is much more to come in the technological displacement of the clock through the network. Future advances will ensure that the relationship between networks and people as users of ICTs will become even more entrenched. As a consequence humans and networks will merge in ways that are positioned to make the régime of the clock ever more marginal to everyday life. Innovations in nanocomputing, biocomputing and quantum computing techniques have challenged both the scale and the very basis on which computing is predicated. These are set to make computing and the role of computers in life even more ubiquitous. Not only ubiquitous, but literally *part* of culture, society and the *physical body*.

We can see this in the development of nanocomputing, which is the construction of computing processes at the nanoscale (nanometer is one billionth of a meter, or one hundred-thousandth the width of a human hair). Working at this level, scientists at IBM have already constructed a transistor that is six nanometers in length that is ten times smaller than previous state-of-the-art transistors (News.nanoapex, 2002). Technicians admit that this scaling down is pushing the limits of what silicon-based transistors are able to do and that they are fast approaching the molecular level where a wholly new approach is required. Such an approach is being developed through research in biocomputing, where computers are able to function like living organisms at the molecular or chemical level, obviating the need for silicon-based technology altogether. And spanning both these developments is the growing research in quantum computing. Here the whole basis on which computing is founded (the binary logic of ones and zeros, on and off) is being challenged. Working at the quantum level, where the classical laws of physics don't hold, engineers have discovered that in quantum computing, binary logic operates *simultaneously* at both one and zero, on and off—at a state they call a 'superposition'—which constitutes a fundamental revolution in the basic nature of information processing. Through these kinds of advances, the cyborg dream of Negroponte (1995) to blend 'bits with atoms' i.e., the fusing of computers with humans, seems to no so far-fetched after all. Indeed, in 2001, Negroponte's brainchild, the MIT Media Lab, with funding from the American National Science Foundation, set up the Center for Bits and Atoms with the explicit aim to 'explore how the content of information relates to its physical representation, from atomic nuclei to global networks' (MIT News, 2001).

It is clear that global networks now constitute an alternative temporal space that now exists alongside the temporal realm of the clock. This means that we live within a double-structure of technological time. One part is that of the clock which is based on the logical and linear time of Newtonian mechanics, and which made possible the imaginings of modernity, Enlightenment science and industrialism. The other is the time of the network which is also based on the strict logic of binary numbers, but grows unplanned, unorganized and chaotically due to its subordination to speed and to the irrational principles of neoliberal competition. It is a temporal divergence that destined to become even more profound and the clock made even more marginal as the binary logic of computing gives way to the 'superposition' logic of quantum mechanics.

For the present, the second Empire of Speed operates within this double-structure of clock and network. In a general sense, time—whether it is clock based or network based—is still about money and money is still very much about time. However, neoliberalism globalization and the revolution in ICTs represent a move beyond modernity and the imaginings of Enlightenment thought. The postmodern network economy and society is oriented towards pure speed in the quest for pure profit. Whereas the clock-based Empire of 'industrial time' did not colonize every realm of life, the networked-based Empire of Speed knows no limit to its expansionist appetite. Work, leisure, culture, polity, society and the physical body have all been gathered up and incorporated into the digital domain, a domain explicitly conceived for reasons of efficiency, flexibility and rapidity. The fragmentary and chaotic nature of time in the network is an unanticipated consequence of rapid computerization and the insertion of abstract values of free market competition. What this portends for economy and society has hardly been considered from the neoliberal perspective. However, a more critical social science which foregrounds the temporal perspective is able to show that this temporal ecology is shot-through with deep social cultural and economic contradictions, what I term *pathologies*—and it is to these we now turn.

CHAPTER FOUR

PATHOLOGIES OF SPEED

These are not normal times.
Sandra Polaski, 2004:1

Précis

In this chapter we will see that in the social world, speed and acceleration are not always the unalloyed benefits that neoliberal ideology portrays them as. Indeed, speed and acceleration have many negative social effects. We look here at what are some preeminent 'pathologies' of this 'bad speed'.

The networked speed economy that has developed over the last quarter century has set in motion what is termed here a global *hyper economy*, which acts as the effect of the post-Fordist mode of production. Speed is of the essence, and through it, the global economy, along with its culture and society more generally, are pulled into the orbit of acceleration—a temporal *hyper now*—wherein humans (error-prone individuals) tend to cope imperfectly. One consequence is a *hyper anxiety*, an epidemic that again has its roots in temporal changes associated with the compulsion towards increasing social acceleration. It is an anxiety that is fundamentally related to our failure to have a sufficient comprehension of what the future might have in store for us. Hyper anxiety, it is argued, is an actual clinical pathology that can have its origins in an unspecific apprehension about the future and what it may bring. What this chapter will show is that the temporal dimension to this problem is real but it is not sufficiently understood in both the social and health sciences where the tendency has been to view anxiety as an individualized pathology.

Emanating from the general pathology of speed is our relationship to information. We consume ever-increasing amounts of it, and at a faster rate. This has led inevitably to changes in the ways that we learn about the world, within the education system, in the media and in everyday life more generally. To cope with more and more information, with the increasing signs and symbols that pulse through the network society—much of which we feel we need to act upon—we develop the habit of what I term 'abbreviated thinking'. This is to think and act quickly because the pressure of social acceleration compels one to do so. As a result, the time needed to think critically and reflectively, the time needed to understand an issue or problem or subject, is reduced through the combination of information overload and speed.

Good speed and the bad

Speed has been good for humans. In many ways it has been very good indeed. Recognition of the power of speed and our ability to manipulate it to our own purpose, distinguishes us from other animals. As humans we have developed an understanding of the physics of speed that enabled us to soar into the air above the clouds, to go yet faster to achieve the escape velocity which opened the way to outer space, and faster still again to aim information-gathering probes beyond our solar system to extend our knowledge of the fundamental nature of the universe. Moreover, humans have found innumerable ways through technological development, to harness the physics of speed to create ways of doing things faster, a humanly-produced speed with which to manipulate and transform our surroundings, our cultures and economies, to become the most successful and dominant species to inhabit the Earth. Today, speed has given us material abundance, and has given us health and wealth and potential beyond our imaginings. On a surface level of comprehension, speed is an unmixed good, and is the very essence of our contemporary world, connecting people and processes and shrinking time and space into proportions that seem 'manageable', thus rendering the physical world as malleable and conquerable.

We needed speed to survive and to flourish as a species. However, at a deeper level of understanding, speed in our postmodern society has become more than the technological 'fetish' that Neil Postman (1993:19) called it. Speed has developed into a craving—a social, physical and psychological addiction, where the faster we go the more is the need to go faster. There is nothing new in this; it's just that we have largely failed to analyze the psychosocial effects of a generalized and accelerating networked-based speed. Possibly this failure stems from our historical connections with speed that tended to emphasize its positive and sometimes thrilling aspects. For example, in early-Victorian literature the intoxicating and exhilarating nature of speed had been recognized at least a century before Marinetti elevated it to a particularly violent aesthetic. In 1834 in his novel *Rookwood*, W.H. Ainsworth wrote an idealistic paean to the English Highwayman Dick Turpin, who had been hanged for his crimes almost a hundred years previously. Ainsworth versified rapturously on Turpin riding his famous horse Black Bess:

> Hall, cot, tree, tower, glade, mead, waste or woodland, are seen, passed, left behind and vanish as in a dream. Motion is scarcely perceptible—it is Impetus! Volition! The horse and her rider are driven forward, as it were, by self-accelerated speed. A hamlet is visible in the moonlight. It is scarcely discovered ere the flints sparkle beneath the mare's hoofs. A moment's clatter on the stones, and it is left behind... He was gone, like a meteor, almost as soon as he appeared.

The elation we feel as our body hurtles through space stems in part from the potential *physical danger* that is inherent in speed. Ainsworth's doggerel speaks clearly to this—both of the speed necessary to escape the pursuing authorities, and of the elation of the ride itself. For us, to travel fast in a car or an airplane (or even ride a horse) is do to something that feels almost illicit; somehow to flout the laws of nature and exceed the natural capabilities of our unaided bodies. People can be 'addicted' to driving cars at top speed, or taking extreme risks when flying an airplane or riding a powerful motorcycle. We can be both thrilled and terrified by the experience. Speed is double-edged, then. It is either 'good' or 'bad'. Goldman provides a useful rule-of-thumb for this notion when he argues that 'Good speed is controlled speed. Bad speed is chaotic, debilitating or uncontrolled' (2004:28). For example, there is the 'good' speed of the escape velocity of space travel that could unlock the secrets of the universe; or of the jet that will take us to the other side of the world to be reunited with family and friends; or of the microwave oven that promises to free up time by rapidly doing certain tasks in the kitchen. And then there is the 'bad' speed of the train crash; or the accelerated production line that the factory worker must keep up with; or of the three hundred unread emails clogging up your high-speed computer, and which every hour delivers more. Such forms of speed, good and bad, intersect our lives on a day-to-day basis, and constitute the speed-essence of our postmodernity. However, the contradictions that inhabit the dark side of speed, the latent pathologies of speed, or the 'bad' speeds that can cause frustration, anxiety, depression, disorientation and forgetfulness (to name just a few), are on the increase. They constitute a generalized and connected *lack of control* at the level of the individual and the system that is new and widespread, and this is due in no small part to the joint effects of economic globalization and the revolution in information and communication technologies.

Hyper economy

In his 1999 book *Hyperculture: The Human Cost of Speed*, which we looked at earlier, Stephen Bertman argues that 'Supported by a network of instantaneous communications, our culture has been transformed into a "synchronous society," a nationally and globally integrated culture in which the prime and unchallenged directive is to keep up with change' (p. 1). Bertman overstates the case somewhat. A 'synchronous society' would be a society where everything operates or flows at the same speed. This is clearly not the case in our digital, networked world where 'network time' functions across an open-ended continuum of speed, from the relatively slow, to the hyper-fast. Moreover to argue that there is an 'unchallenged directive' presumes a coordinated intent or a coherent human agency that is in control. A key factor in this analysis of the networked society in the context of neoliberal globalization is that there is essentially no one in control. The global economy is energized by the abstract demands of profit and the need for ever more increases in the pace of production, transaction, and consumption across the whole of social and economic life. Nevertheless, within this reality of social and economic acceleration, Bertman is right to argue that we are driven to 'adapt' to its constantly increasing tempo. We are driven to this through what Marx called the 'dull compulsion of economic relations' which registers as the need to keep a job at the level of the individual, or the chasing after that ever-ephemeral 'competitive advantage' at the level of the business entity. Bertman is also right to suggest that the human cost is massive, a transformation in our networked societies that has witnessed a deterioration from '…an acute state to a chronic condition'. He goes on to argue that 'Its effects…are not only psychological; they have become ethical and moral as well' (1999).

Recall that Bertman wrote this in 1998. At that time the Internet was only just beginning to boom and to spread its tentacles far and wide (Hassan, 2004). Cell phones were still fairly exotic, and the ubiquity of computing throughout economy and society had still to reach saturation point. Moreover PDAs, VoIP phone and video calls, file-swapping, chat-rooms, instant messaging, webcasting, podcasting, Web 2.0

technologies, on-line universities, 'weightless' corporations and so on, were either non-existent, on the drawing board, or still in their infancy. The networked society has grown massively since even that recent point in its history. For example, the Pew Internet & American Life Project could write in 2005 that:

> A decade after browsers came into popular use, the Internet has reached into—and in some cases, reshaped—just about every important sphere of modern life. It has changed the way we inform ourselves, amuse ourselves, care for ourselves, educate ourselves, work, shop, bank, pray and stay in touch.

This is a bald fact. What the statement does not convey is that the Internet, by invading 'every important [and not so important] realm of modern life', has also extended speed and acceleration into every nook and cranny of our existence. Neoliberalism and digital technology create the surface imagery of a shiny world of efficiency and convenience, but the 'chronic condition' that is the reality of networked speed is daily taking a human toll. Network speed is creating a hyper economy and society that is as familiar to people as it is abstract and alien, so powerless do many of us feel under the influence of its unstinting pace.

The hyper economy is another term for neoliberal economic globalization. Its speed mania is actualized in, for example, the global stock, currency and investment markets. These are driven by networked, high-speed computer systems working within a global human ecology that operates on a 24/7 basis. The stock market is primarily an exercise in number crunching. Its logic is based on speculation—guessing the prices and the likely financial/economic consequences of shifts in the marketplace. This process is governed by a very inexact science. It comprises the heady admixture of a market economy that has many altering and contingent features such as buying and selling; environmental factors such as climate change or natural catastrophes; social aspects such as greed, thrift, recklessness, criminality, prudence, government policy, war, poverty, consumption, etc.; and to this is added the very human propensity to refuse to function as the rational actor that neoliberal theory puts at the center of its view of the marketplace. These all swirl in a complex totality that defies ordering and predictability in any efficient way. But this is what the stock market tries to do: traders, investors, consultants and so on, stake jobs and careers and savings on being able to make sense out of this. And to make matters worse (and even less understandable and controllable) it tries to do this as quickly as possible, much of it driven by computer programs that operate on a global-wide 24/7 basis.

This high-speed guessing-game can involve the movement over three trillion dollars a day in currency trading alone (Forex.com, 2008). Where time is money, speed is of prime importance. Where prices fluctuate constantly, profits are made on the speed and timeliness of the transaction. The volatility and the turbulence that comes from institutionalized speculation and risk may not matter but for the fact these markets have a direct effect on the reality of daily life: seemingly safe jobs can evaporate overnight, the payments on a mortgage can be manageable one day and impossible the next and the savings that were put aside for your retirement can be suddenly lost or massively devalued.

Such speed is inherently unstable, and so the acceleration that pulses through the global networked economy necessarily brings volatility with it. The social effects of this may take differing forms within differing contexts, but the fundamental effect of *instability* is ineradicable. This fact generates one of the key pathologies of the second Empire of Speed. Instability is a factor that is both constant and generalized. For example, illegal Chinese workers toiling in the kitchens of London restaurants operate on a differing spatial and temporal context than Forex traders in Wall Street, but the instability and insecurity that both groups feel emanate from the same interconnected and deeply rooted processes of neoliberal economic globalization and the revolution in ICTs. One can argue, of course, that volatility, unpredictability and insecurity have been a feature of human existence for thousands of years. However, what is unique about our own time is that these are rooted in human creations and come from a singular economic, ideological and technological dynamic. It is a dynamic that is innately destructive in regard to our relationship with time, space, the past, the future, the social world and the environment that houses it. The London dishwasher and New York share-trader *are indeed* connected and affected by global acceleration in ways and with an intensity that previous generations were free from. Their fates, moreover, rest not in their own hands, nor even the hands of their bosses, but are decided through the calculus of an abstract conjunction of free markets, rapid technological development and the persistent compulsion to accelerate further and faster.

Hyper now

The economic and technical striving towards temporal immediacy produces the effect of a *hyper now*, or what Robins and Webster term an 'eternal present'. For them 'What this means is that global society is being subordinated to a rational and standardized time'—the time of the network (1999:235). To exist in a hyper now temporality necessarily weakens or dissolves our links with the past and the future. For example, history and the past take time to research and to reflect on for their patterns and possible lessons, but time compression and acceleration render these less relevant to a logic that clamors for immediacy. As regards the future, Robins and Webster state that 'A major achievement of the capitalist imaginary has been a colonization of the future—and that means the colonization of possibility' (1999:235). With the future increasingly closed off in terms of its diversity of potentials, the present functions as a kind of temporal prison. Paul Virilio (1997:25) sees this as a kind of real-time society, what he terms a 'live (live-coverage) society' with 'no future and no past' and where alternatives to what immediately confront us seem difficult to imagine or impossible to achieve. However, the future, when one does have the time to think about it, appears inevitably as a continuation of the present because in this postmodern context the Enlightenment and its future-oriented ideals of progress, planning, security and stability are being broken up and scattered to the whirlwinds of rapid change.

When past and future cease to be strong anchors on which one's subjectivity is tied, then the sense of one's life as an ongoing and coherent narrative project becomes more difficult to imagine and sustain. Social acceleration acts to lock us into a temporal postmodernity that is present-oriented. As Lorenzo Simpson (1995:144) puts it, this process:

> ...can be characterized as an attitude toward time or an experience of time that...places emphasis on maximum intensity in time, not the living in time that would be a form of praxis, but a more passive fascination or playing...The result is a flashing pointillism, a lived experience as a series of disconnected intensities. Not being able to commit to a future or to take the past seriously, the postmodernist makes do with the present.

Within the hyper now there is the feeling of speed without the exhilaration or danger that an Ainsworth or a Marinetti might exalt in. These exciting characteristics are supplanted by a *disorientation* that stems from the intensity of the eternal present, intensity brought about by the demands on the individual acting as a node within the network of flows. The physicality of hurtling through space as a referent for speed is being replaced by the reality of the individual who is relatively still in space (at a computer, on a cell phone, at home, walking, driving, and so on) but existing within a fast-moving and connected cyberspace. Ironically, then, within the accelerating network of information flows we are trapped by a growing bodily inertia, a remote action mode whereby ICTs virtually *extend* us into cyberspace. It is a space that envelops the individual in what Gleick (1999:217–225) sees as a 'net of efficiency', and in the service of efficiency we are pushed, cajoled and pressured into new psychic contortions in the search for this evanescent state.

To exist within the hyper now it is not even necessary to be connected, physically, by cable or wirelessly. The network society is so pervasive that it creates its own meta-environment that surrounds us and has its effects whether we are 'connected' in the strict sense or not. Simply being part of the global digital environment means that we cannot escape its demands or its accelerative presence for long. For example the bill that arrives in our letterbox the day after you used your credit card heightens the sense of temporal immediacy. Similarly when shopping in, say, a British supermarket for local tomatoes, the compression of space and time becomes literally perceptible when we discover that the only tomatoes available are (cheaper) Israeli ones that were picked less than twenty-four hours before. Or one can enroll for a university degree course that once took three years, but now for the sake of efficiency and 'convenience', can now be offered (for a fee) for completion in two (Hassan, 2003). The point is that the sense of a contracting global space and time are not strictly limited to the inhabitants of cyberspace, but is something that affects us all in one way or another through the general networking of the world and its socio-economic processes.

The difficulty is how to respond, how to understand the stresses and strains of the hyper now. Speed alters perspective. This means that the actual interrelation of things, images or processes can disappear, become blurred or unconnected altogether. This blurring and disconnecting are forms of abstraction through which social acceleration masks the reality of the world and their actual connections. Raymond Williams acutely observed the changes in perspective through the acceleration of visual information in the 1970s in his book *Technology and Cultural Forms*. Writing about his experiences of television watching in the USA, he noted that:

> I began watching a film and had at first some difficulty in adjusting to a much greater frequency of commercial 'breaks'... two other films, which were due to be shown on the same channel on other nights, began to be inserted as trailers. A crime in San Francisco (the subject of the original film) began to operate in an extraordinary counterpoint not only with the deodorant and cereal commercials but with a romance in Paris and the eruption of a pre-historic monster who laid waste New York. I can still not be sure what I took from the whole flow. I believe I registered some incidents as happening in the wrong film, and some characters in the commercials as involved in the film episodes, in what came to seem—for all the occasional bizarre disparities—a single irresponsible flow of images and feelings (1974:21–22).

Williams describes the process of watching a television screen and being disoriented by it, albeit in a way that may seem quaint in our media-saturated culture. Today television is only one aspect of the crowding of our retinas and ears with fast-flowing images and sounds. Moreover, television was and still is monological. Try as it might, by getting viewers to ring a toll-free number or visit a programme website, or text a celebrity, or vote yes or no on almost any question, television remains essentially a one-way traffic flow where you can always 'tune out'—and no-one really cares. In the network society, however, interconnectivity at high-speed is a dialogical process, especially since the advent of so-called 'collaborative' Web 2.0 technologies that began to emerge after 2004. Responses are expected if not demanded. People are always waiting for an answer. Nowadays you 'tune out' on pain of disconnection. And not being connected can cost you your job, or the chance of a better job, or you are denied the narcissistic knowledge that you are riding the cusp of the latest technological wave. So you 'choose' to stay connected and run the risk of disorientation, perplexity and confusion: did I send that email to John in New York? That voicemail asking for the report to be attached by email by tomorrow: who was it that sent it again? Why doesn't Alice answer my text messages?

Disorientation and immediacy, the twin markers of the hyper now, also feed directly into a sense of pervasive crisis, of a feeling that we cannot cope and that society is now permanently 'on the edge' (Hutton & Giddens, 2001). This can be crises in the world 'out there' in the economy, or in the seemingly unending cycles of conflict, the war against terrorism, the scourges of famine and diseases such as HIV AIDS; or it can be the innumerable micro-crises that insert themselves into our daily personal existence, such as the loss of a job, a spike in interest rates, or the latest foreshortened deadline that has rushed unexpectedly to meet us. As Virilio noted in his *Speed and Politics* 'the immediacy of information immediately creates the crisis' (1986:143). Everything has to be done quickly and when it is done quickly, mistakes invariably are made and crisis ensues; and when it is not done quickly enough then crisis will result anyway. Such 'bad' or 'uncontrolled' speed presents us with a no-win situation.

Increasingly unable to prioritize (to time manage) effectively, and unable to appreciate the bigger picture (the complex totality of the globalized, networked economy) we fall behind in what is demanded of us and get anxious about it. Multi-tasking (significantly, a term derived from the austere and asocial province of computer science) has become the neoliberal buzzword for trying to think about, and physically achieve, more than one thing at a time—and all in the name of efficiency. Ida Sabelis quotes the response of a CEO from a major corporation to her question on time management in the context of a high-speed economy. The executive quipped that 'The art of leaving things out…is the answer to the need or wish for things to happen faster' (2004:294). Of course the answer sounds somewhat glib, and reads like making a virtue—or an 'art'—out of instrumental necessity. But there is no doubt that 'leaving things out' because of time pressure is, for most people, a source of cognitive and emotional anxiety. Coping strategies naturally emerge—and leaving things out is one of them. However, most of us are more immediately answerable to family, to bosses, customers, or colleagues for our inactions. And so we try to multi-task. We get a faster computer, a more powerful organizer, a more multi-functional cell phone, a wireless network connection with a longer range, more wireless access points at more locations, and so on. We do all this to keep connected when out of the workplace and to enable us to take work home at weekends or at night. Empirical work on the connections between ICTs and the enabling of more tasks in less time indicates that 'multitasking "adds" on average more than seven hours to each day, totaling an addition of more than fifty hours to the average week' (Kenyon, 2008:291). The willingness to be so flexible and so productive reflects what most people know—that that employment is now deeply insecure. For increasing numbers of us, the prospect of a non-renewal of the next employment contract serves to focus the mind on where the priorities lie. And the priorities, distressingly, appear to us to be with yet more connectivity, to try to achieve more through multi-tasking and to be more efficient in more realms of life without realizing fully that the more we try to do the less we can do it effectively or efficiently.

A daily life that is packed with the requirements of multi-tasking is a daily life that is inhabited by ever more fragile non-mediated connections, that contain increasingly faint narratives and traceable pathways within them. Long-term projects, personal commitments, times for reflection, for planning, for future orientation that would develop a sense of historical continuity, simply cannot satisfactorily gel in a life dominated by speed and networking and constantly having to juggle with an increasing number of tasks. It is impossible to adequately reconcile these inherently asynchronous temporalities, and this is in no small part because, as Hartmut Rosa wrote, the '...ideal of the autonomous and reflective leading of a life requires adopting long-term commitments which bestow a sense of direction, priority, and "narratability" to life' (2003:20). The fact that we are increasingly denied these age-old human aspirations make the hyper now a condition that surpasses the non-specific *angst* of modernity. It mutates into a social pathology that manifests as a form of anxiety that has clear links to postmodern social acceleration.

Hyper anxiety

We have seen that speed has become a form of economic and social addiction. But what of the psychological addiction? Like other addictions, speed is one addiction that has withdrawal symptoms and severe side effects, some of which are only now becoming problematized in the social sciences. The disorientation and sense of immediacy that emanate from social acceleration feed directly into a generalized pathology of *anxiety*.

Now, the extent and locus and nature of anxiety in our modern (and postmodern age) have long been a moot point within the social sciences. The following discussion draws extensively from the work of Iain Wilkinson, a sociologist of health, because his expert focus of anxiety as a social phenomena helps illuminate the issues that are of concern in this book. Wilkinson makes the observation that '...anxieties expressed by sociologists over public issues such as the environmental crisis, the resurgence of nationalism and historical events such as the demise of communism, are by no means proportional to those encountered in the wider population' (1999:446). 'Reasons for anxiety' he further contends, 'do not necessarily raise feelings of anxiety' (ibid.).

On the face of it, this would seem to complicate things. However, a useful way to move through this problem is to begin at the beginning and ask: what is anxiety? A further complication we immediately encounter, as Wilkinson notes, is that anxiety can only be experienced subjectively 'from the inside' and therefore 'cannot be observed or measured from the outside as an object in itself' (p. 453). This is part of the problem that social science and social theory more particularly has had with ascribing 'anxiety' to 'an age' (Dunant and Porter, 1996) or to particular social processes (Giddens, 1990). Accordingly, some speculative points of insight are offered here that may be useful in ascribing an economic, technological and temporal locus to the undoubted plethora of forms of anxiety that characterize postmodern society. This is a ubiquity that is hardly in doubt and is evidenced in the numbers of individuals stirred to seek professional help. It is evidenced too in the growing demand for the palliative of prescription drugs for anxiety, stress and depression supplied by a $400 billion pharmaceutical industry that is ready and willing to assist (Angell, 2004).

Freud argued that anxiety is related to *fear.* He maintained that fear is a reaction to a perceived situation of danger that can be identified, and that anxiety stems from a situation (a danger) that we do not know or fully perceive (1979:284). As Wilkinson notes from this Freudian perspective that the 'condition of anxiety appears to take place when people lack a *sufficient understanding* of the threatening situations in which they find themselves' (1999:450). This is a profoundly temporal condition: *immediate* danger produces fear, a situation that may not be fully understood, but is at least recognized as a danger to the self. As theologian Paul Tillich (1980:44) observed, fears 'have a definite object which can be faced, analyzed, attacked [or] endured'. Anxiety, however, is more diffuse and is usually accompanied by undefinability and the inability to establish a sufficiently clear idea of the cause of the apprehensive feelings we experience. It is, in the temporal context, an *anxiety about the future*, as opposed to a fear of what actually confronts us in the here and now. Wilkinson touches on this key issue when he writes that '...the condition of anxiety is vitally related to our failure to provide ourselves with a sufficient comprehension of a perceptibly hazardous future [and that] anxiety entails being emotionally disturbed and psychologically traumatized by the knowledge of our ignorance of future events which appear to place us under some kind of threat' (p. 452).

This condition has been a fact of social life since at least the beginnings of the industrial revolution, when constant change was itself constituted as a social fact of life. However, in the network society, anxiety about the future has become much more generalized and the future much more foreshortened. In the context of a hyper now, where speed flows through a network-driven economy into all facets of life, we are given much less time to deal with all demands adequately or to fully understand the nature of the tasks that confront us. The unknown produces anxiety. Risks or uncertainties that are known can be confronted and planned for, whereas a limited knowledge of risks, or just an unspecific apprehension about the future and what it may contain can be psychologically debilitating.

We do not experience fear in this 'constant present' because the 'threats' and the 'risks' are multifaceted, they are not a 'definite object'. They are linked to other threats that make them blur and move too fast for rational comprehension and understanding—and thus avoiding. This lack of knowledge is equated with a lack of control over our situation. The future thus rises as a dark shadow on an ever-closing horizon where we imagine all manner of vaguely comprehended dangers lurking in wait for us. Accordingly in the economy and society of the 'constant present' increasing numbers of people are 'traumatically aware that [they] lack a sufficient understanding of why [they] are feeling so disturbed' (Wilkinson, 1999:452).

To take one example: in our postmodern society jobs are tenuous. 'These are not normal times' argues Sandra Polaski (2004) in her study 'Job Anxiety is Real—and it's Global' which looks at the effects of ICTs in the creation of a postmodern labour market. Structural changes in the economy have created a situation where, Polaski observes, 'trust [between employees and employers] is under strain' (2004:1). Workers are increasingly employed on contracts that give little or no guarantees of stability or predictability. A high-speed economy demands the prevalence of such jobs, the more the better. That 'people fear for their jobs' is an often-used cliché in contemporary postmodernity, where the 'jobs for life' maxim has gone the way of old-style Fordist industry. Perhaps, however, it would be more accurate to say that people are anxious about a very uncertain (and deeply opaque) future, where not having a job is only one catastrophe that they feel can befall them.

Another example: personal debt has reached record levels in major economic regions such as North America, parts of Western Europe and Australasia. This debt represents money that has been eagerly lent to people by banks who need to compete with other banks for their custom. Speed again is an important factor here. If one bank will approve a loan in 24 hours, a rival feels compelled to better this by doing it faster—over the phone or the Internet. Competition means that the banks cannot (or will not) slow the process down, sufficient to allow them to properly and responsibly manage the money they lend. Debtors take this increased risk on themselves. We do it partly because we live in a consumption-dependent society that it no longer oriented so much to the future. And we do it partly due to the blandishments of saturation advertising that trumpet the joys of unlimited and pain free consumption. It is at the very least arguable that an effect of this 'easy money' is that we tend to sublimate thoughts of paying it back at that future time which is an already hazy vista. However, no matter how hard most of us try to repress the thought of owing growing amounts of money to the bank, there is also that angst-ridden part in us that thinks (reminded by a solicitous or demanding letter from the bank): 'how *am* I going to pay for this house', or car, or renovation? The anxiety-generating bad dream became a reality for millions of people after the 'sub-prime credit crunch' of 2008 that began in the USA when the extent of unsafe lending practices finally became apparent. The temporality of finance capital was hitched to an unsustainable present and a radically discounted future. The inevitable eventually happened and personal bankruptcies, mortgage foreclosures and a serious 'credit crunch' began to vibrate around the world through a deeply interconnected global banking system (Turner, 2008).

Anxiety can flow from counter-intuitive sources, too. In the rich Western economies and in the advanced countries of North and South East Asia, consumer cultures are replete with every convenience and every service to the extent that anxiety can stem from having *too much choice*. What to buy when advertising bombards with advice? What to eat or drink or buy when science tells us that our lifestyle might be life threatening? What lifestyle, indeed, should I have? Tillich argues that anxiety is expressed in a 'loss of direction' and a 'lack of intentionality' (1980:44). Speed is a core element of these society-wide pathologies, and anxiety, stress, depression are its social and cultural manifestations. But as Wilkinson notes, anxiety is subjective, something that only the individual feels fully in its intensity and its range. Social scientists thus find difficulty in measuring or theorizing social anxiety and identifying, however speculatively, its root causes. However we can turn this problem around and look at it from another angle. The economy, globalization, job insecurity, levels of indebtedness, the institutionalization of the short-term perspective, the rapid turnover in product cycles, lower real wages, turbulent stock markets and so on, are *objective* social conditions that *can* be measured. These are social forces that give rise to an interconnected world that is far more complex and far less comprehensible that it was, say, a generation ago. This complexity, this opacity, this denial of the future and valorizing of the present, constitute a temporal context (speed-filled and future empty) within which the subjective experience of trauma and anxiety are given their locus and expression. People inhabit this context and it is the measurable spread of uncertainty, complexity and lack of coherence and rationality within the postmodern neoliberal and digital economy that allow the necessary social conditions for anxiety to broaden into a social pathology.

Anxiety as the cultural condition of the speed economy

Anxiety takes many forms and individuals cope with it (or try to cope with it) in many differing ways: through medication, natural remedies, yoga, meditation, alcohol, violence, shopping, anger—or simply shrugging their shoulders and attempting 'get over it' as best they can. Yet if we take the Freud/Tillich/Wilkinson approach as a guide, then a degree of *increased understanding* of daily life and its complexities, as well as some feeling of control over the future, would alleviate much of the anxiety that currently thrives in our lives and in society more generally. The argument here is that we need to understand why we are anxious to get to the root of the problem and then work towards alleviating or solving it. The crucial question we need to ask, however, is whether our levels of understanding are *sufficient* to help us come to grips with the problem—or whether, as I will now discuss, we are simply looking in the wrong places and identifying the wrong causes.

Frank Furedi, in his book *Therapy Culture* (2004) tracks the massive increases in the 'cultural condition' and 'therapeutic language' that we know as anxiety, stress, depression, and so on, over the last twenty years. It is a condition that we easily recognize and a language that we now readily use in order to describe our apprehension with the postmodern world and its ways. Furedi provides empirical evidence that indicates that anxiety and stress are not limited to working-age adults who need to keep up with the pace of change in the speed economy. Children as young as 9 or 10, Furedi reports, now express feelings of being 'stressed out' or anxious, and primary school pupils in at one institution in Liverpool, England '...are being offered aromatherapy, foot and hand massages, as well as lavender-soaked tissues to help reduce stress and aggression' (2004:1). The author goes on to note a UK survey which indicated that 53 per cent of university students have 'anxiety at a pathological level', and Furedi cites another by the British Association for Counseling and Psychotherapy which found that ten per cent of students seeking professional help are 'already suicidal' (p. 8).

There is a significant cultural, social and economic problem, then. Anxiety reaches into every demographic with a scope and an intensity that is historically unparalleled. Furedi underscores this prevalence with reference to a survey of instances of certain social-psychological terms being brought into general currency:

> ...the term 'syndrome' was entirely absent from the pages of American law journals during the 1950s, 1960s and 1970s. Yet by 1985, the word 'syndrome' appears in 86 articles, in 1998 in 114 articles and by 1990 in 146 articles. In one month alone in 1993, more than 1000 articles in periodicals and newspapers used the term. In Britain, the growth of a therapeutic vocabulary is equally striking. Most people would easily recognize words that were virtually unknown and unheard of by the public in the 1970s in the early 1990s. Even in the 1980s, people had never heard of terms like generalized anxiety disorder (being worried), social anxiety disorder (being shy), social phobia (being really shy), or free-floating anxiety (not knowing what you are worried about) (2004:2).

By labeling the problem do we understand anxiety in ways that would empower us and help us deal with it more effectively, to stop its social and cultural spread? To be sure, there has been no want of trying to deal with the problem and 'treat' it at the level of the individual. Since the 1980s a huge anxiety industry has sprouted in most Western economies. Counseling and counselors permeate the whole spectrum of government and non-government sectors. The police, army, schools and universities as well as government departments all now offer counseling services to staff and students, as do most corporations in the private sector. Yet the 'industry' and the 'problem' continue to burgeon. Writing in the late-1990s James Nolan observed that 'The monumental increase in the psychologization of modern life is...evident in the fact that there are more therapists than librarians, fire-fighters, or mail carriers in the

United States, and twice as many therapists as dentists or pharmacists' (1998:8). Counseling has been 'institutionalized' in Britain, Furedi notes, and permeates 'all levels of education and the penal system' (2004:10). The spread and growth of the counseling, therapeutic and also the pharmaceutical industries would seem to indicate that we are no closer to understanding the source of the problem of anxiety as a cultural and social condition, and therefore no closer to halting its spread as an individual (or individualized pathology).

In trying to assess and critique this development, Furedi constructs his thesis around the idea that the pervasive 'therapy culture' is about imposing 'a new conformity' through the exploitation and management of people's emotions. In this schema, businesses in the private sector supply counseling as a 'powerful instrument for maintaining managerial control' (2004:99)—a rationale we can presumably extrapolate to public sector service industries as well. Some serious questions need to be raised at this point. Is the need to 'control' people and press people into 'conformity' really the motive force here—and are the present-day mechanics of social control as nefarious as Furedi claims? It is difficult to imagine that managers (or anyone in business) sit around and discuss such tactics of control and then actually implement them through the offer of counseling. Conformity may be an *effect*, but it is not the rationale behind the spread of anxiety and uncertainty in the world. Furedi's book is useful in that he correctly identifies a serious social and cultural problem—that of a mushrooming culture of anxiety and therapy that has reached extraordinary levels. But to identify the *cause* as emanating from a new form of class control does not seem to be supported by any convincing evidence.

For causes that would logically cohere with the overarching thesis of social acceleration we need to look at the chronology of the rise of the 'psychologization of modern life'. It is significant, for example, that the growth of anxiety and uncertainty throughout society (and the industries through which to identify and treat it) correlates with the emergence of neoliberalism and the revolution in information and communication technologies. It has the same temporal trajectory, in other words, as the speeding up of society. An effect is that the interrelated dynamics of neoliberalism and ICTs has tended to atomize society into what Ulrich Beck and Elizabeth Beck-Gernsheim (2002:33) describe as an 'isolation of individuals within homogenized social groups'. Within this growing privatization of the self, traditional social support networks become frayed and fragmented as the ideology of individualism strengthens correspondingly. In this 'runaway world', such feelings of isolation and powerlessness, together with the experience of uncertainty and inability to project a recognizable and safe future for themselves, their families and their loved ones can lead to anxieties, to neuroses and to all the new kinds of social-psychological pathologies that we have identified and labeled in the last quarter of the 20th century.

We must locate this dynamic squarely within the bleak realm of economics, of free markets, and the accompanying compulsion to make individuals and systems as efficient and productive as possible. The institutionalization of neoliberalism and digital networks in the service of profit has created the context in which the pathology of depression can flourish. As has been remarked before, this is not a conscious process formed in the heads of the elites and then forced on to others in order to control them and make them conform. Those wielding social, economic and political power, like the rest of us, are in thrall to speed and are forced to short-term solutions for immediate problems. The pressures of time for managers and CEOs can be even greater and so mostly they are simply not able to indulge in grand thought-out plans for long-term domination and consolidation of power. We are dealing with a deep structural process marked by ever-dwindling forms of institutional/human control.

Moreover, this system *feeds on* the pathologies it creates, exploiting them as by-products of profit seeking. Neoliberal doctrine teaches that almost anything can be a market 'opportunity' to be actualized as profit. And so the market has 'captured' and colonized human anxiety as an opening for new business, creating, effectively, an industry out of it as well as the general social-psychological fall-out of economic restructuring and the creation of a networked society that have occurred since the late-1970s. It does this through the same processes that have created an 'education industry' out of the deregulation of universities (Hassan, 2003); or the 'consulting industry' created out of the evisceration of levels of management in the fashioning of so-called 'weightless' corporations to compete in the 21th century (Rifkin, 2001). Furedi makes this point himself when he quotes psychotherapist Nick Totton as describing counseling training as a 'pyramid selling scheme' whose principal effect has been to create a 'huge increase in its clients' (2004:10).

These colonization and commodification processes take place at the level of an almost autonomous *system*. Individual actors, investors, health professionals and so on, work *within* this overarching logic. People accept, by and large, that this is how the world is. They internalize and accept the concept that anxiety and stress are personal problems, and that alternatives that point, at least it part, to the present system as being responsible for these pathologies are only utopian fantasies (Anderson, 2004). Moreover, the speed economy and the neoliberal economic environment that creates it, is not the 'root of all evil' in the spread of anxiety across all demographics. What occurs is that neoliberal capitalism as an abstract system creates the socio-economic and political *context* that greatly *amplifies* already existing social pathologies. Social pathologies such as anxiety in turn are explained by transforming them into an industry around which a legitimating discourse arises. Furedi's point on the disconcerting trend of children being habitually 'stressed-out', for example, is well taken as an indicator of how far the problem of social anxiety has reached into society. However in the following example, we can see how at the level of the abstract system—as opposed to the alleged control-freakery of managers—anxiety and uncertainty become amplified, commodified and industrialized in the service of speed.

In February 2005 the London *Independent* newspaper reported in its 'Money' section on the British company YoungBiz, which runs courses in personal finance, entrepreneurship and marketing. The average age of the 40 students who were currently enrolled in their programme was between 13 and 15. 'Financial education', these young people were informed, is today's 'hot topic' with a 'generation of people... faced with the prospect of working through to their grave, or accepting a much poorer standard of living in their retirement' (Daley, 2005). Speed, temporality, anxiety and insecurity are all bound up here in a neat little YoungBiz prospectus. Children are brought in early to the heart of the speed economy in order to understand that in an uncertain and predatory world you can never be too young to gain 'financial literacy'. The future is a forbidding realm and young people must be trained in the ideology of self-reliance because if you don't look after yourself, no one else is going to. The effect is that the time span of childhood and its natural innocence is compressed yet further as the harsh 'realities' of being born in this world at this time are made known at an early stage in their life. Of course one can only speculate on the levels of anxiety that drive parents to send their children to such 'training courses' on top of their regular schoolwork, or on the levels of 'free-floating anxiety' that these children themselves experience, both as result of their parents worries for them, and their own worries about their future and their ability to provide for it.

YoungBiz and the many similar businesses that thrive in such a social and economic context are not out to 'control' young people and make them 'conform' to a right-wing, pro-business worldview through making them even more 'stressed-out'. These may indeed be *effects*—however the people who run such companies work within a specific setting that gives them a specific perspective on the way the world actually is. By truncating their childhood and making children into self-seeking and self-preserving individualists who view the world apprehensively and through a profit and loss calculus, YoungBiz teachers and the parents who send their children there, no doubt imagine they are giving them the best possible start in life. What they create, however, at the level of the individual, is the raw potential to become (in a very short time) atomized actors within a fast-paced and uncertain world who perceive their social experience through a preoccupation with the self and their own feelings of failure and inadequacy in the face of an inability to cope. The result, as Beck and Beck-Gernsheim argue 'is that *social problems* are increasingly perceived in terms of psychological dispositions: as personal inadequacies, guilt feelings, anxieties, conflicts and neuroses' (2002:33) (italics added). Ultimately, then, in the neoliberal economy of speed we increasingly misrecognize the source of our feelings of anxiety—we believe it to be *us* that have the problems and therefore *us* who need to change, instead of the reality being a particular construction of society's processes.

Abbreviated thinking

Perhaps the most insidious pathology of speed—and it is one that is related to the pervasiveness of anxiety—is that as society becomes more complex and fast, the less we understand it, and the more we inhabit it at a surface level. As Dean Kuipers observes '...constantly treading water at the surface of change has consequences. Deeper historical currents flow beneath us [and] we don't see them as clearly as we should' (2000:12). Speed breeds the cognitive tendency towards what I term 'abbreviated thinking' and its effects permeate society and shape how we understand society and interact with it.

We are told that the network society is also an 'information society' and a 'knowledge society' (Bell, 1973; Tapscott, 1996). These terms are used positively in much of the general literature and typically connote the excitement of technological development and the tremendous opportunities that it is supposed to herald (Rheingold, 1993; see also Hassan, 2008 for a review of this literature). Information and knowledge are indeed central elements of the network society. However, the contexts in which these are generated, comprehended and disseminated, work to produce social and cultural effects that *actually diminish* our understanding of the world and our individual and collective ability to shape its future directions.

Again speed and acceleration are key to an understanding the emergence of this new relationship to information and knowledge and to the growing resort to abbreviated thinking that these produce. What is meant by the term 'abbreviated thinking'? The argument over the changing dynamics of knowledge production within neoliberal, postmodern society has been made in detail elsewhere (Hassan, 2003:109–142) and there is no need to recount it here except in a broad-brush way. Suffice to say that throughout the period of modernity the production of knowledge emerged out of and helped to strengthen, two modes of thinking—critical and instrumental. Both are a form of reason that have roots in Enlightenment thought and reaches still further back into Greek antiquity. Both were vital to the construction of modern societies.

Critical thinking, or critical reason, functions dialectically and is able not only to question and add to other forms of knowledge but is also able to interrogate itself and the ontological foundations of its own knowledge production. Critical thinking, working in its optimal environment, would generate its own time and take the necessary time appropriate for dealing with the question or problem that is being confronted. On the other hand, instrumental thinking, or instrumental reason, is essentially non-reflective; it is goal-oriented and much more concerned with producing results and outcomes, be they techno-scientific or politico-economic. Instrumental thinking is concerned, moreover, with achieving results as quickly as possible and with as little fuss as possible, to get straight to the heart of the question or problem and produce a (usually) technological 'solution' for it.

During the course of modernity neither critical nor instrumental reason functioned optimally, or were given totally free rein to exploit the potential they contain. Both tended to work in some sort of interaction—and necessarily so. This is because if humans thought and acted on the basis of critical reason alone, then hardly anything would get done. It would be a pure 'thought world'; a world constructed largely in the mind, a world where nothing is certain, where there are no *a priori* assumptions, and where other ideas and processes need to be taken into account, endlessly. Instrumental reason, on the other hand, as Ronald Barnett (1997:91) puts it:

> ...takes the world largely as given and attempts to find means of living ever more productively and efficiently in it...instrumentalism works within a horizon of ontological assumptions. The world is objectified: the task is that of securing effects in it and on it. Objects, events, situations, technologies, knowledges and persons are valued so long as they have a use value.

The interaction of both these modes of thinking fitted well with the project of modernity. Indeed it was central to modernity. The application of critical thinking was able to project a possible world of freedom and of justice in human affairs, whereas instrumental thinking acted on the concrete reality of the world 'productively and efficiently' through the development and application of scientific and technological forms of knowledge. Each could complement the other: instrumental thinking could jolt critical thinking out of its inertia; and the application of critical reason, in its turn, could act as a check on the baleful effects of instrumentalism—on the 'totally administered world' that Herbert Marcuse saw as the end-point of instrumental reason, or the 'iron cage' of rationality that Max Weber saw as an innate tendency within modern society.

It is vital to understand the inherent temporal dispositions of these ways of thinking about the world and how well (or how poorly) they are suited to the current social, economic, technological and temporal order. By foregrounding the time perspective we can see that instrumental thinking with its emphases on productivity and efficiency, of identifiable goals and 'ontological assumptions' regarding the way the world is, meshes easily with the priorities of speed. Critical reason, on the other hand, is antithetical to speed, to rushing around, to coming quickly to conclusions and then acting rapidly (and often thoughtlessly) on them.

It is only with the advent of neoliberal globalization and the information technology revolution (and the foregrounding of the temporal perspective with which to analyse these dynamics) that the vital disjuncture between the critical and instrumental dialectic becomes apparent. This postmodernity, with its hegemonic emphasis on speed, efficiency and productivity is creating the context for what Ken Goldberg identifies as 'a recalibration in our definition of knowledge' (2001:3). What this means is that what we consider as valuable and desirable in terms of knowledge has shifted decisively towards instrumental forms, replacing the complex and age-old interaction between the critical and the instrumental that had characterized the building of modernity. Jean-François Lyotard noticed this shift—and within this specific speed context—as early as the late 1970s in his book *The Postmodern Condition: A Report on Knowledge* (1979). In it he writes that a momentous shift is now underway, one that is set to change the 'state' of knowledge into a lop-sided domination by instrumental forms of knowledge and thinking. According to Lyotard, the universities and the information technology revolution hold a central responsibility in this. He writes that 'that the status of knowledge is altered as societies enter what is known as the postindustrial age and cultures enter what is known as the postmodern age' (p. 3). Lyotard argues that 'technological transformations' in computerization, information storage in data banks, etc., 'can be expected to have a considerable impact on knowledge' (pp. 3–4). He goes on to argue that 'the miniaturization and commercialization of machines is already changing the way in which learning is acquired, classified, made available, and exploited' (p. 4). Anticipating the arrival of the commercialization of the learning institutions, Lyotard notes that the old notion that knowledge and pedagogy are inextricably linked has been replaced by a new view of knowledge as a *commodity*, and as a consequence, teaching and learning has become an alienated and alienating process:

> Knowledge is now produced in order to be sold; it is and will be consumed in order to be valorized in a new process of production: in both cases, the goal is exchange.

This commercialization of knowledge reaches into the heart of the university and through its power Lyotard predicts a shift in the whole system of organized learning:

> It is not hard to visualize learning circulating along the same lines as money, instead of for its 'educational' value or political (administrative, diplomatic, military) importance; the pertinent distinction would no longer be between knowledge and ignorance, but rather, as is the case with money, between 'payment knowledge' and 'investment knowledge'—in other words, between units of knowledge exchanged in a daily maintenance framework (the reconstitution of the workforce, 'survival' versus funds of knowledge dedicated to optimizing the performance of a project (p. 6).

Computerization is the technological means of speed and neoliberalism its ideological collaborator. Beginning in a systematic way during the 1980s, these dynamics began to transform Western universities, the main institutional and cultural basis for the creation of instrumental and critical forms of knowledge (Delanty, 2001). Under neoliberalism the university was increasingly cast as a 'business enterprise'. It was thrown by market forces into the business of making money (indeed it *had* to find alternative sources of funding as government subventions were gradually cut back). It was compelled to learn to swim or drown, like any other business, by grasping market 'opportunities'. Today universities are driven to conform to the laws of the market to an extent that is historically without precedent. Acting as businesses they must produce the forms of knowledge that the market (and industry) demands (Marginson & Considine, 2000; Hassan, 2003:69–125). Operating with an eye fixed permanently on the 'bottom line', university departments and their mushrooming layers of non-academic management have discovered certain harsh economic realities. For example, industries (and increasingly students themselves) tend not to see the immediate relevance of, say, the inculcation of a habit critical thinking through a liberal arts degree, or of a three or four year quest for knowledge for its own sake, or the study of medieval poetry, or politics, or pointillism in visual arts, and so on. In the today's instrumentalized climate of higher learning these are seen to offer little in the way of income for the university or employment opportunities for the student. Both institution and student are obliged to focus on the reading of marketplace signals to see what 'skills' industry might need, and then develop as soon as possible (before a competing university gets in first) the optimal mix of appropriate vocational courses to exploit this 'opportunity'. Under such ideological and economic circumstances there was only one possible outcome for the production and dissemination of knowledge. It is an outcome that Barnett (1997:92) succinctly articulates, describing how 'knowledge' transmutes into 'information' in the service of an instrumentalized 'information economy':

> Critical thought cannot be construed just as a form of individual action or mental state. We live in a knowledge society. This *is* the case, but more to the point are the changing forms of knowledge. Humanities give way to science; small-scale forms of knowledge production give way to large-scale forms; knowledge for its own sake gives way to applied knowledge; pure inquiry gives way to problem-solving *in situ*; prepositional knowledge gives way to or at least is supplanted by experiential knowing; and ways of knowing give way to sheer information.

The dominance of applied instrumental knowledge—and to think in an abbreviated way—constitutes what Ulrich Beck (2004:131) terms the new 'epistemological referent' of the age of neoliberal globalization. Through ubiquitous and fast information technologies, knowledge is increasingly produced, exchanged and disseminated through a digital nexus which shapes its form and content to align more easily with the needs of the economy, with profit and with efficiency.

The obverse side of this process is that critical reasoning and the forms of knowledge that it generates have become devalued and are being consigned to the margins of social relevance. The shift to the hegemony of instrumental knowledge production and the production of ever-increasing volumes of information that we must now contend with (at ever-increasing speeds) means that we now are likely to conceive, comprehend and consider in an abbreviated fashion. For increasing numbers of us there is simply no time to delve deeply into forms of knowledge and modes of thinking that have no direct economic relevance. Where do we find the time to analyze and discuss the ethics of abortion or euthanasia? How to allocate the time to be politically aware, sufficient to make informed decisions about the complex and complicated political developments that make such a difference to our lives? Who will spare us the time to sit down and help us develop knowledge of political economy adequate to make the national and international economic situation at least some way comprehensible? Like Sabelis's (2002) CEO, we 'leave things out' because we have no choice; there is no time. We think about the world in an abbreviated way, leaving gaps in our knowledge and awareness that become wider and darker as social acceleration continues.

In the universities subjects such as history, philosophy, economics, literature and anthropology stagnate or die (or commercialize and degrade), whilst the study of accountancy, computer programming, business and tourism studies, marketing and psychology grow and thrive. In this postmodern society, where the market and its 'laws' are preponderant, where open-ended speed is seen as 'natural' and desirable, and where the past and future contract into a constant present, it is no wonder that anxiety abounds. If human cultures and societies have trained themselves to be less critically aware, then the world and our place in it inevitably seems more opaque. Social, economic and political dynamics continue, of course, but the causes of crises, of economic boom and bust, of political and ethnic strife become more and more indecipherable for most of us. And so in life, surprises continually spring on us individually and collectively, and we are left ill equipped to properly comprehend their nature or deal with their effects.

As was indicated at the beginning of this section, abbreviated thinking is akin to Kuipers' metaphor of 'constantly treading water at the surface of change'. Deeper historical currents do indeed flow powerfully beneath us but we can barely feel them or appreciate the fact that we are being driven along by them, because we, individually and as a society, lack the grounded and anchored 'truths' that modernity used to furnish (for better or for worse); the narratives of past and future and progress and the feeling of a 'centre' that would constitute our place in the world. This is not to say that modernity with its interaction of critical and instrumental reason uncovered immutable 'truths', but it did produce forms of knowledge that could act at least as provisional truths, as the 'metanarratives' that gave coherence to the life project of the individual and the collective. These truths held and were able to guide as well as anchor so long as there was a consensus regarding their validity. The move to our current postmodern sensibility has pulled the supports out from these structures. Acting as a pathology, social acceleration drives us to thinking in an abbreviated fashion, and causes us to view the wider world in that way. The deeply insidious aspect of this process, however, is that such thinking functions in a vicious cycle. It is like senile dementia: the worse it gets, the less we are aware of it.

Humans are essentially cultural and social beings. We create meaning and we create cultural forms out of any context. Digital speed and neoliberal society likewise produces a culture of meaning, a culture that is growing in vibrancy and in scope to the extent that it is an 'industry' and an 'economy'. However, this too is now based on the technologies of computing and networking and has expressions in every realm from art to politics and from 'new media' to leisure. It goes under the general rubric of 'cyberculture' and is the subject of the next chapter.

CHAPTER FIVE

CYBERCULTURE: A CULTURE OF SPEED

> *Cyberculture is the culture that has emerged, or is emerging, from the use of computer networks for communication, entertainment and business*
> (Wikipedia *definition, 2008*)

Précis

As an ecology—as an environment that increasingly dominates the space and times that individuals and societies live within—the network society 'naturally' produces its own cultures. Such cultures are formed through the digitally produced signs and symbols that surround us; through the streams of binary information that flow at an ever-accelerating rate through the Internet; and through the networked totality that comprises the network society. This is called cyberculture. But what is it? What does it reflect? And where are people placed within the network society in respect of their cultural autonomy? This chapter tries to answer these questions by first looking briefly at the 'pre-history of cyberculture'. Here the assumptions that underpinned the development of modern computer science are discussed. It is shown that theorist/engineers such as Norbert Wiener with his work on cybernetics, and J.C.R. Licklider's conceptually-related work on 'mancomputer symbiosis', transformed the discipline of computer science and provided the building blocks upon which the Internet and the network society would be constructed. The vision of these men was to build a computer-driven society where humans are in control and where, to paraphrase Licklider, men will set the goals and computers will do the work.

The chapter goes on to argue that because of the effects of the unique networking capacities of computers, we need to look again at the dynamics of technological determinism. To a significant degree the shape and form of cyberspace (and therefore cyberculture) is 'determined' by the instrumental and commercial logic of the network system, as it is currently architectured. The effect is that 'cyberculture' is in fact a 'speed driven culture' and reflects the narrow logic of the neoliberalized global economy. An example of this is discussed through 'game culture', a preeminent form of cyberculture that has been argued by some to be antithetical to the mores of capitalism and the ideology of conformity. It is shown, however, that 'game culture' as well as cyberculture more generally, reflects the values and imperatives of a networked, global, and neoliberalized capitalism almost exactly.

> *The mythology of cyberspace is preferred over its sociology.*
> Kevin Robins (1996)

On a general level, and emerging from the early influential books on the subject, from media and from the global advertising of ICT corporations, the neologism 'cyberculture' connotes all that is good and affirmative and creative. It suggests a whole new world of possibility stemming from networked information technologies, a world where people communicate freely and effectively, bringing together multicultural perspectives into a global whole where mutual understanding at the level of the global can create informed and secure communities at the local level. *Community* has been a continuous trope in discussion on cyberculture. In 1993 Howard Rheingold published an influential book called *The Virtual Community* in which the growth of network-driven cyberculture was analyzed in untiringly positive terms. Rheingold thought it useful (because we are dealing with real people) to view cyberspace and cyberculture in biological terms. The metaphor he chose to use (perhaps unfortunately) was that of the organism floating in the Petri dish. He writes that:

> In terms of the way the whole system is propagating and evolving, think of cyberspace as a social Petri dish, the Net as the agar medium, and virtual communities, in all their diversity, as the colonies of microorganisms that grow in Petri dishes. Each of the small colonies of microorganisms—the communities on the Net—is a social experiment that nobody planned but that is happening nevertheless (p. 10).

The network, for Rheingold, is a virtual space that mirrors the 'real' world in terms of its community-creating potential, where people can 'do just about everything people do in real life' (p. 9). The evolving network and its communities is diverse and vibrant, he argues; it is not a monolith, but functions more a 'like an ecosystem of subcultures, some frivolous, others serious' (ibid.). The 'ecosystem' of subcultures that the network generates, as Rheingold and many others have argued, is overwhelmingly positive (apart from 'deviant' behavior such as cyber rape, or cyber stalking or the growing forms of cyber crime). Cyberspace is held up as the first culture-creating context where space and time become increasingly irrelevant and where technology is used creatively and where technology is put at the service of ordinary people. In this Rheingoldian view people are no longer 'slaves to the machine' as they had been under the ponderous and innovation-crippling phase of Fordist modernity. As at the dawn of the Age of Enlightenment, so the argument goes, everything now seems possible again, once the shackles of the previous era had been thrown off. Above all, this view sees cyberculture and the network communities from which they emerge, as processes that bring with it coherence and a degree of internal control; a 'true grassroots use of technology' as Rheingold puts it, in the building of a myriad of potential and actual cybercultures (p. 9).

There are obvious truths in elements of this largely uncritical perspective. Network communities and cybercultures do indeed permeate virtual space and time. All kinds of communities of interest create and share cultural meanings and symbols from which forms of cyberculture develops. From swapping online notes on pigeon-fancying with like-minded individuals from around the world, to arthritis sufferers who email each other ideas on existing treatments or alternative therapies, people build a cyberculture of a kind which would fit into the affirmative framework outlined by Rheingold and others. Moreover, there is self-evident value in these communities in that they serve a 'community' purpose and develop a 'culture' that in many ways reflects its members and their interests.

What is proposed here, however, is to analyze more fully the preponderant trends in the development of cybercultures in the network society; to look at the cultures that are *mass* and *global* and where people spend more and more of their time in specific cybercultural contexts that are more reflective of the (usually ignored) underlying philosophical and technological premises on which computer science is built. In other words I wish to develop a critique based on what Darren Tofts has called the 'pre-history' of cyberculture (1997). This is a valuable pre-history where we find the ideas, the ideologies and processes that made possible the network society. And *contra* the almost universally positive view of cyberculture, the incorporation of its own pre-history into the analysis throws a wholly different light on how culture is produced through cyberspace technologies, and what kind of culture it is. For that we need to go back in time and look at the beginning of what is generally understood as cyberculture.

A short pre-history of cyberculture

To understand the basis of cyberculture we need to grasp the ideas, the politics and above all the technologies that underscore its evolution. The technology of which I speak is of course the computer, and I have already discussed the effects of the elemental logic of computation that is premised on binary numbers. However, the revolutionary turn that computer science took in the middle of the 20th century was guided by specific assumptions about what computers represented, and by powerful intellectual figures who gave these theories, these assumptions, concrete reality.

Evolving out of the basic number-crunching role that characterized their early development, computers came to be seen in the post-war era as having tremendously powerful potential as *systems of control*. It was conjectured that humans would be able to use computer systems as a means with which to transform their environment, orienting modernity more comprehensively toward plannable futures. This was a vision that was given much state support in the USA in the context the Cold War. Military strategists, for example, could see the benefits of computer-driven 'command and control' coordination that cut out as much as possible the 'human error factor' when dealing with vastly complex atomic and nuclear weapons systems (Edwards, 1995). Largesse from successive administrations obsessed with the 'Soviet threat' spurred a generation of scientists and engineers from Ivy League universities who began to develop the rudiments of what would become the science of 'cybernetics' and lay the intellectual and technical groundwork for the Internet.

The term cybernetics has its etymological roots in the Greek word *kybern*, meaning 'to steer' or 'to pilot' (McHoul, 1998). It was a term that referred to maritime navigation and has obvious connotations to control and to movement through space by means of technology (ships, rigging, oars, etc.). In the 1940s, Norbert Wiener, an engineer, mathematician, philosopher of Logic and scientist at the Massachusetts Institute of Technology (MIT) utilized the word to prefix the term cybernetics, a school of thought he pioneered. Cybernetics, for Wiener, was a theory of both control and communication through flows of information. Fundamental to cybernetic theory was the claim that it applied equally to humans and machines. Feedback mechanisms were a central link to this working in practice. Information feedback, Wiener hypothesized, would generate automatic processes between the human and the machine. Automatic doors, for example, work on this principle. A person approaching such a door alerts a sensor that 'understands' this human action and triggers a 'response' which is the activation of a mechanism that causes the door to open. Human and machine create this 'automatic' action through mutual interface—through feedback. Cybernetics was, then, at its very inception, a mode of thought that was deeply linked to the concepts of 'command and control' and the search to discover ways in which environments could be engineered and manipulated through human-computer interaction.

Wiener, like Isaac Newton, whom we encountered earlier, saw the world and its reality as being made explicable by number. Mathematics was key to understanding how the natural world functioned. And so digital computers, being based on binary logic, were for him analogous to the functioning of the natural world. Wiener saw computers as machines that mirrored what humans were in their essence. In his 1948 book *The Human Use of Human Beings: Cybernetics and Society* Wiener was unequivocal about this link:

> ...the [human] nervous system corresponds to the theory of those machines that consist in a sequence of switching devices in which the opening of a later switch depends on the action of precise combinations of earlier switches leading into it, which open at the same time. This all-or-none machine is called a *digital machine*. It has great advantages for the most varied problems of communication and control (1948:11).

Equating humans with computers through a system of cybernetics was a powerful hypothesis buttressing the development of computer networking and the systems of communication that would eventually become the Internet and the network society. For example J.C.R. Licklider, an acknowledged intellectual father of the Internet, wrote a highly influential paper in 1960 called 'Man-Computer Symbiosis'. In his *Introduction* he writes that:

> Man-computer symbiosis is an expected development in cooperative interaction between men and electronic computers. It will involve very close coupling between the human and the electronic members of the partnership. (...) In the symbiotic partnership, men will set the goals, formulate the hypotheses, determine the criteria and perform the evaluations. Computing machines will do the routinizable work that must be done to prepare the way for insights and decisions in technical and scientific thinking. Preliminary analyses indicate that the symbiotic partnership will perform intellectual operations much more effectively that man alone can perform them.

At the beginning of his essay Licklider seems to alternate between viewing computers as *partners* with humans in the symbiotic relationship, to seeing them as mere tools that *serve* to make our lives easier by performing the routine 'intellectual operations' that will extend our 'technical and scientific' capabilities. He goes on, however, to argue that symbiosis is in fact a necessary and unavoidable phase in human development, without which human life would be much degraded. In other words he implies that we *depend* on computers—a very different proposition from there being a partnership or symbiosis in existence. To illustrate this he uses the metaphor of a variety of fig tree that is only able to reproduce through the pollination of a single insect, the *Blastophaga grossorun*. Both species he observes are 'heavily interdependent' on each other, and indeed each would die without a deep and direct symbiosis. Licklider's point is that humans need computers if they are to progress as humans. We need them because as they are based on the logic of numbers, powerful computers are able, in the Newtonian view, to decode the mysteries of how the world and the universe function. And if we can direct and manipulate these celestial tools, we can then unlock modernity's potential, penetrate the essence of reality, and render it open to direction and control.

In Wiener and Licklider we see two significant and mutually compatible theses on the nature of computing and the nature of humans vis-à-vis computing. Wiener argues that to be fully modern is to depend on computing power, and Licklider suggests that computing also reflects our human essence. These perspectives are at the core of post war computer science; and the work of these towering intellectuals in information systems and cybernetics contain basic assumptions regarding people and computer systems that flow directly into the creation of our networked society. Both theorists view humans as fundamentally *processors of information* in a world whose reality becomes apparent only through the interface with numbers. As Theodore Roszak wrote in his book *The Cult of Information*: '…in perfecting feedback and the means of rapid data manipulation, the science of cybernetics was gaining a deeper understanding of life itself as being, at its core, the processing of information' (p. 39). The central challenge for humans aspiring to full modernity, so the logic runs, is to develop information processing machines that increase in their sophistication, speed and capacity. The more powerful computers become, the more we realize our inner essence, and the more we are able to understand the world around us.

The logic underpinning Wiener and Licklider's theories seems to be that a relationship with a specific form of computing is a necessary phase of human development; however, they do not phrase it as a *dependency*, which would be the logical corollary. Instead, both are careful to inscribe a *hierarchy* where humans remain in control, 'formulating' and 'determining' the uses of computers and 'evaluating' the success or otherwise of their operations. This would be a benign form of control where machines are developed to the point where autonomous 'thinking' and action could be performed. The ultimate goal of this symbiosis in Licklider's paper of 1960, for example, was the realization of a world where computing had moved away from the solution of straightforward 'preformulated problems' to one where they would be more involved in the *formulation* of problems. The idea would be '...to bring computing machines into the processes of thinking...' itself (pp. 10–11). This is an early conceptualization of artificial intelligence (AI) that, for Licklider, would have its most immediate use in military 'command and control' systems in Cold War planning (Edwards, 1995:ch8). In a more general sense, this symbiosis (not dependency) would be expressed through the fusing of intelligent humans with intelligent computers to fashion a mighty tool that would be a crowning achievement of modernity. Our reliance on 'simple' tools would have finally been transcended. Through AI we would have created intelligent 'partners' to help us achieve our highest aspirations. This new phase in human-technology relationship would bring computing into the very centre of what it is to be human and modern. This would be the fulfillment of an age-old dream that stemmed from Newton and Leibniz in the period of Enlightenment, through to Charles Babbage and Alan Turing in the 19th and 20th centuries, respectively.

Effectively functioning AI systems, in symbiosis with humans or on their own, are still some ways from existence. But the inner logic of computing that would drive them are very much part of our daily lives. With a passion and comprehensiveness that Wiener and Licklider would commend, computerization is today seen in business, in government and in society more generally, as the very pinnacle of 'control' and efficiency that puts the user 'in charge' of 'intelligent systems'. From personal cell phones to vast corporate systems, computers are now almost universally conceived of as the way of the future. For those who believe in the project of modernity, computing stays true to the faith placed in science and technology by the Enlightenment founding fathers. In other words, we can still imagine computer technology developers, legislators, businesspeople and users themselves as 'steersmen' who still hold firmly to the wheel of progress on course towards a technologically sublime future. And as 'steersmen' we bring the day ever closer to the realization of the Negropontean (1995) vision of fusing 'bits with atoms' in the creation of a seamless interaction between machines and humans. This will be a benevolent cyborg world where deep technological mediation blurs forever the point at which humans end and technology begins. It was a projection that Gordon Pask took to its conceptual limits in his 1982 book *Microman: Living and Growing With Computers*, where he argued that 'mancomputer symbiosis' would in fact create a form of human immortality, ideas that hitherto were the preserve of the sci-fi genre of 'cyberpunk' literary fiction (Pask, 1984). In Pask's imaginary, not only would cultural artifacts and records be preserved, but also individual human minds and their personalities. This would be the culmination of the project of modernity that the *philosophes* of the 18th and 19th century could scarcely have imagined, but would have doubtless approved. And the 20th century theories of Wiener and Licklider would have been instrumental in taking the project of modernity to its zenith.

At its very roots, then, the search for optimal systems of *control* is what computer theory is about, and by extension, so too is the networked society that is its supposed expression. Control is the central philosophy and driving force. Control over our surroundings, control over the means of production in society, control over our lives and control over our futures. This is the computer dreamscape and the metaphysic of binary logic. Whether most of us give a thought or a care to the 'essence' of computing is something else. The ideology of neoliberalism and its obsession with computers as the ultimate 'solution in search of a problem' has dulled our collective critical senses (Postman, 1993). Notwithstanding the lack of reflective debate on its causes and effects, cybersociety is a growing *reality* and the cybercultures it spawns are undeniable. Accordingly the profound ontological questions that this reality poses will not disappear into the cyberspace ether. The concrete actuality of cybersociety continues to emerge ineluctably through '...the proliferation of electronic media, and the increased automation of everyday services, facilities and activities [that] have meant that the character of everyday life is becoming less proximate and more virtual' (Tofts, 1997:14). And so we *need* to think and we need to care because notwithstanding its philosophy of control, cybersociety is a society that reflects our computer *dependency* and *lack* of control.

A theme throughout this book has been that the impetus behind the (unintentional) development of the network society and the (very intentional) 'need for speed' that has been its primary characteristic, is the nexus between neoliberal globalization and the ICT revolution. As Michael Benedict argues, cyberculture is an expression of this political and economic dynamic and represents '...a new stage, a new and irresistible development of human culture and business under the sign of technology' (1993:1). The political economy rationale has focused on the economic, political and the ideological processes that have created our postmodern and virtual speed-space, wherein time and space flow at ever-increasing rates of acceleration that we humans must try to synchronize with. This is a *temporal political economy*, in other words. What makes these economic, ideological and political abstractions real is, as Benedict reminds us, technology. This is a technologically determined space and time that creates its own cultures, its cybercultures. The question of *technological determinism* thus becomes a salient and unavoidable issue where questions of control are concerned. To what extent is cyberspace a space (and time) that humans have real control over? And to what extent are the cultures, the meanings, symbols and practices that emerge from this virtual space and time, reflective of free and open forms of communication?

Technologically determined cyberspace

In social theory there are two main ways of looking at how technologies function in modern society. First, and usually the most easily dispensed with in academic debate, is the theory of technological determinism. As the term suggests, it is a theory which gives a defining role to technology as the shaper of society and its individuals. It is a technology-led theory of social change wherein people and institutions simply conform to technological developments. Classic examples are McLuhan's 'the medium is the message' aphorism from his *Understanding Media* of 1964, and Alvin Toffler's 1970 book *Future Shock* in which he wrote of '...the shattering stress and disorientation that we induce in individuals by subjecting them to too much change in too short a time' (p. 330).

Apart from these well-known examples, such theorising (and theorists) tend to be thin on the ground in the social sciences. This is because insufficient agency is percieved to be given to *social* forces, such as culture, economy and politics. Indeed, to argue a theory of 'technological determinism' is to often court ridicule for being naive, or if not naive, then in some way vaguely authoritarian.

On the other hand, a *social* determinism argues that it is society—its people, and institutions such as politics, government and the economy—which functions in elaborate, dialectical ways to shape technology which then in turn reshapes society. The theory holds that people can 'adopt or reject' technologies though our ability to act as autonomous 'consuming agents'. A typical account comes from Hannah Rippin (2005) who explains that:

> Politics, economics, culture and organization are crucial to the invention, design, adoption and implementation of technology. The ways in which technologies are required and used by society are driven by market forces, and the adoption or rejection of technologies are shaped by social action. The design and production of the technology will be shaped by technologists, but the *ultimate choice* lies with the consuming agents of society. Technological artifacts, although introduced into society, are not forced on it (italics added).

Variations of Rippin's description of social determinism or 'co-determination' have become the acceptable face of theories of technological development. It is a theory that satisfies a deep-set tendency within the social sciences to emphasize the social and the 'ultimate' power that it wields. As the now-classic textbook on the subject edited Mackenzie and Wajcman maintains:

> We will take technological change as a given, as an independent factor, and think through our social actions as a range of (more or less) passive responses. If, alternatively, we focus on the effect of society on technology, then technology ceases to be an independent factor. Our technology becomes, like our economy or political system, an aspect of the way we live socially (1985:5).

In some ways this line appears as unarguable because it gives a shaping role to both society and technology. It underscores the eminently sensible idea that it would be difficult to think of circumstances where people have ever been completely passive in respect of technology—or of anything else. Agency, resistance, or the thinking of alternatives, is part of what it is to be human in the world (Williams, 1979:252). However, it is also important to recognize that in the debates over technological or social determinism, talk is of technology in the abstract sense, or when examples are given they almost always in the context of *individual technologies*. Rippin, in her otherwise useful article on the cell phone, is emblematic of this intellectual tendency, and cites the personal stereo—and in the context of the wider argument—the cell phone itself, as exemplars of 'co-determination' (Rippin, 2005).

As may be seen in this short explication of theories of technological development, the question of *control* is central. And 'co-determination', i.e., that society 'ultimately' shapes technologies, is generally seen to be the most serviceable explanation. Looked at from the perspective of networked speed, however, it may be useful to develop another way to approach this question. In an almost throwaway line in a posting to the online Media Ecology listserve, media theorist Rob Blechman (Media Ecology, 2005) nonetheless dramatically opens up the lens to wide-angle in respect of the debates over technology and society when he asks:

> Why does technological determinism have such a bad rep? While I always hold out for individual freedom and choice and I don't discount the impact of public policy, I believe that across large periods of time and large populations a case can be made for the determining effects of technology. Recent attention paid to the effects of natural climate change on the rise and fall of civilizations and the potential of 'man-made' greenhouse gases to bring an end to our civilization should bring discussion of technological determinism into the forefront of the debate.

Systems of techology, in other words, can have systemic social (and therefore individual-level) effects. The countless technologies that combine to contribute to global warming are having a determining effect on nature, and on the actions of society and individuals within it. What is at issue here is that we should no longer take technologies (either as 'evolved' from earlier versions, or from other 'branches' of technology) as discrete analytical forms. This is especially the case with computer-based technologies. Accordingly, the cell phone today (if we use that example) is not simply the particularly clever progeny of the first cell phone developed by Bell Labs in the 1940s. Neither is it a technology that mysteriously fired the popular imagination in the 1990s to became incredibly prevalent. The cell phone is a part of the wider revolution in ICTs that have transformed society since the late 1970s. It is, alternatively, part of a *digital logic*, a systemic force that transcends the power and potential effects of any single technology. Cell phones by themselves do not 'determine' users' behaviours, but the immensely more potent and all-encompassing network society of which it is part, does. It therefore requires a category shift in thinking about technology to understand the significance of this. The problem needs to be approached from the perspective of *systems of technology* and accept that generalized, networked computerization has given this technology new social and cultural power. Michael Heim expresses this point in his analysis of Heidegger when he writes that:

> ... technology enters the inmost recesses of human existence, transforming the way we know and think and will. Technology is, in essence, a mode of human existence, and we could not appreciate its mental infiltrations until the computer became a major cultural phenomenon (1993:61).

McLuhan's alleged crime of technological determinism may be seen in a rather more mitigating light if we change the focus of the argument from individual technological artefacts to meta-systems based on the logic of 'command and control'. McLuhan's 'medium is the message' may be rephrased here without loss of signification as 'context informs content'. In our own time this means that the context (networked ICT systems) generates the immense flow of content (information) that goes to make up cybersociety. What Robins and Webster term the 'transformative power of information' (1999:75) *works on* society and its individidials, orienting them towards instrumental ends as they spend more and more of their time in ever-accumulating, ever-accelerating data-streams. Their vulnerability stems from the neoliberal restructuring of society's institutions of collectivity and power that had characterized modernity. People now become the weak link in relationship to this techno-system. Networked society pulls us towards it through the 'dull compulsion of economics', and we are growing dependent because we *need* to be connected to live and work and to be part of the 'normal' mainstream of networked life.

Speed, the core of the entire process, makes us yet more susceptible to the determinating force of the neoliberal network society. And what is accelerating, of course, is digital technology and its singular product—information. It surrounds us and we enter into its flows through our laptops, desktops, PDAs. It is a gadget-driven and networked-powered febrility that Lash sees as a constant '... movement from interface with auto and cell phone to aeroplane to television to pager to the streaming-enabling baseline software in our TV set-top box' (2002:10). Speed and volume mean that we are hardly in a position to make sense of all these hypermediated flows of information as they bombard us, much less control them effectively. For example, in 2005 a US survey found (unsurprisingly) that our lives are mediated as never before—to the point that 'the average American spends more time using [networkable] media devices—television, radio, iPods and cell phones—than any other activity while awake' (Ransford, 2005). And so we yield to them. We do this consciously or unconsciously, to the point where the effect of ubiquitous computing and the increasing time-space compression emating from open-ended techological speed begins to consume our now-vulnerable subjectivity, our faculties of cognition and our modernist perspective on the world. Herbert A Simon, cognitive scientist, developer of AI systems, and 1986 Nobel Laureate for economics pondered the nature of accumulated computer-generated information in society:

> What information consumes is rather obvious: it consumes the attention of its recipients. Hence, a wealth of information creates a poverty of attention and a need to allocate that attention efficiently among the overabundance of information sources that might consume it (Simon, 1971:40–41).

To 'efficiently' deal with this 'overabundance' is easier said than done in the context of the time-squeezed network society. Efficiency, in the neoliberal lexicon is a never-complete technical-rational process that demands always the application of yet more computing power and speed. And as the application of computers generates yet more information at faster speeds, 'overabundance' and what Virilio (1997) calls 'information gridlock' become an even bigger problem. An alternative to this, as Thomas Hylland Eriksen argues, is to use ICTs less, to try to limit their use and gain more control over one's time (2000). The problem here is that being out of the information loop is now hardly an option if one wishes to pay the rent or mortgage, have a decently-paid job or simply remain in the mainstream of economic and social life. Furthermore, opting out merely avoids issues of social control through recourse to individual denial—which is no answer at all. And so we invariably opt for the 'technical solution' in the illusionistic search for autonomy through using the faster processor, the more multifunctional personal organizer, the more powerful laptop, and so on.

The tremendous determining power that digital information posesses stems from the valorization of speed. We saw how during the the development of early modernity time was quantified, commodified and marshalled as an economic resource. This same logic applies to information in the network society. Time and speed are central to the potency of information. As Scott Lash observes, during the phase of modernity 'Use-value and exchange-value had a past and a present'. However, in our high-speed postmodern society, *information* is the primary use-and exchange-value. Here 'Information-value is ephemeral. It is immediate. Information-value has no past, no future: no space for reflection and reasoned argument' (2002:144). For Lash, this is 'out of control' information, creating a '*dis*information society'. It is information that creates a volatile global economy and the deep-level systematicization of unintended consequences in a meta-context where, as Lash puts it, there is a diminished 'logical and analytical ordering' force in society (ibid.).

Speed-driven digital information has profound metaphysical consequences, too, affecting our capacity to understand the reality of the world. Lash goes on to observe that our network reality is contructed through instrumentalized information that is 'shot through with the facticity of the particular' to the detriment of any universals. It is an information-generated reality comprised of '...the pure empirical with the disappearance of the transcendental' (144–145).

This is the system-level process that inculcates the abbreviated thinking that I see as a social pathology, or Simon's 'poverty of attention' at the level of the individual. Not only does the blizzard of information make us less able to understand the complexity of the world around us, it compounds further the power of systemic technological determinism. Speed-derived disorientation and the illusion of efficiency that ICTs supposedly bring, create what Robins and Webster see as '...the general sense of acquiescence to innovation' (1999:74). They go on to argue that:

> ...technology, without discernible origins, is something that ordinary people cannot understand. The technology is a mystery, and it remains a mystery even when its technical functions are explained in simplified terms, because its genesis—its social history—is ignored. (...) Without history, the new technologies have become an unstoppable force which, though incomprehensible to natives, is understood sufficiently for them to realize that they must change their whole way of life.

In their different ways McLuhan and Toffler understood the power of technological systems as opposed to technologies analysed at their discrete levels. They understood the power of a generalized techno-logic, even if they could not have foreseen the precise and profound consequences of computerization. McLuhan, for example, understood systems of communication, but misunderstood speed. In his *Laws of Media* he wrote that: 'At electric speed, all forms are pushed to the limits of their potential' (1988:109). By this he presupposes, as does the neoliberal argument, that 'speed is good' and that through it we will be able to realize ourselves as moderns. He did not foresee the detrimental consequences of social acceleration through 'electric speed' that surges as pure information and which is, as Lash puts it, 'out of control'. Above all he did not realize that this systemic speed, this techno-speed, this *ideological* speed, would subjugate people through its valorized and instrumentalized orientation.

Toffler on the other hand more clearly perceived the dangers of a new systemic force in history. He termed it the 'accelerative thrust' and saw in it a new and determining power (1970:402). In his still-readable *Future Shock* he notes with a consciously un-Luddite eye the speed of his 1970s world, a world then on the threshold of yet greater technological and social transformation.

> The speed-up of diffusion, the self-reinforcing character of technological advance, by which each forward step facilitates not one but many additional forward steps, the intimate link-up between technology and social arrangements—all these create a form of psychological pollution, a *seemingly unstoppable* acceleration in the pace of life (p. 388). (my italics)

Our alienation from the history of technology, or more precisely, from the genesis of systems of production and efficiency, makes the speed-up seem 'unstoppable'. Toffler recognizes that the issue is a political one which involves the 'conscious regulation of technological advance' (p. 387). His is not an argument for limiting technology; indeed we need more of it, he insists, but only under greater social control. His 'future shock' is the observation that we 'face an even more dangerous reality: many social ills are less the consequence of oppresive social control than of oppressive lack of control. The horrifying truth is that, so far as much technology is concerned, no one is in charge' (p. 290). Toffler's presentiment lies in the fact that he could see the effects of systemic speed a decade before neoliberalism abnegated much social control over technological development—and left it to market forces instead.

Toffler's approaching 'dangerous reality' is our cybersociety: a meta-context of speed wherein people dwell and try to synchronize their lives to its quickening tempo. From this reality new cultures (cybercultures) emerge and grow. These comprise the living expression of the nexus between technology and social self-identity. Analysis of these 'cybercultures of speed' will conclude this section.

Speed driven information culture

In recent times the subject of cyberculture at the popular level has begun to generate a growing literature. Much of it, as Darren Tofts puts it in his *Prefiguring Cyberculture*, 'addresses the matrix of themes to do with the integration of human life and technology' (2002:2). However, a fairly uncritical enthusiasm still outweighs critical analysis. A growing awareness of what people do in cyberspace and the possibilities it may indicate has prompted a multitude of responses in such realms as art, literature, film and design. Numberless courses have been developed in the universities, hoping to capitalize on its fashionableness with younger generations, and also to seek money-spinning 'synergies' with new 'creative industries' hungry for tech-savvy workers who feel at home in this digitally created space and time (CIRAC, 2005). Indeed the European Graduate School (EGS) in Switzerland is an example of an entire institution devoted to the study of cyberculture. It is, predictably, a 'virtual university' with an optional summer face-to-face programme for Masters and Doctoral students. Again, the effort here as in much cyberculture activity, is oriented towards industry and the commodification of the process. EGS is organized to attract well-paid professionals from the new media industries with the prospect of higher-degree accreditation from a university virtually staffed by a galaxy of cyberculture practice and theory superstars from around the world (e.g. from Tracy Emin and D.J. Spooky, to Georgio Agamben to Zlavoj Zizek) who comprise its (bizarrely diverse) virtual faculty (EGS: 2008).

What is lacking in much of this intellectual and commercial activity is an understanding of, or critical perspective on, what these hugely significant technological and social dynamics represent. What is missing is an analysis that looks at the mechanics of cultural production within cyberspace and the foregrounding of speed and time to the very centre of this human life and technology nexus.

A possibly more productive way of approaching an analysis of cyberculture is to delicately alter the angle of vision by giving the subject a slightly different name. It is an name with the very same meaning and connotations, but carries with it a flexibility that allows more easily for its deconstruction. For the purposes of this thought experiment let us change the term 'cyberculture' to 'techno-anthropology'. Admittedly it does not sound as mellifluous or as cool, but it does allow for some intellectual purchase in the attempt to get some deeper insight and meaning.

We have already looked at the meanings and effects of information technologies, the dynamics underpinning the 'cyber' or 'techno' prefix. But what of the 'anthropology' of cyberspace, of the dynamics of cultural production and reproduction that take place within what poet Richard Brautigan (1967) termed the 'cybernetic ecology' of our postmodern era?

A motivation that deeply pervades the production of cultural practices in human societies is that of *control* (Adam, 2004:143–148). The need to feel able to have some degree of power over our relationships with other humans and with the environments that surround us is a primal survival instinct—even if we do not always see it in such terms. As we have seen in the works of Wiener and Licklider, with their emphasis on control systems and human-computer symbiosis, their *oeuvre* can be seen as contributing to a technological expression of this entrenched tendency. If the need to control (communication, order and understanding) is an impetus behind cultural production, we still need to dig a little deeper to ask: what then is culture?

In his *The Interpretation of Cultures* anthropologist Clifford Geertz maintained that the concept of culture 'is essentially a semiotic one'. In other words, it is through the production of signs and symbols that humans derive meaning and understanding (1993:5). He goes on to note that '...man (sic) is an animal suspended in webs of significance he himself has spun, I take culture to be those webs...' (p. 5). To stretch this metaphor a little further: the spider *controls* its environment through the spinning of the web which delivers it food, just as humans attempt to control their environment by surrounding themselves in 'webs' of meaning through the signs, symbols, values (and the technologies) they create to allow them to survive and thrive. Importantly, this is not a passive process. We can augment Geertz's idea of the production of culture as being geared towards the construction of frameworks of meaning by utilizing the concept of *action* as motivating the actual mechanics of culture production. As Erik Kline Silverman puts it in his treatment of Geertz's work: '...structures of culture guide the individual's actions, but only through action do structures and culture become real' (1992:124).

We can see that this control action in cultural production is present at the metaphysical level in the human construction of religions and cosmologies that ordered and structured the world from its earliest social beginnings (Adam, 2004). For example, ideas of time and space have been deeply implicated in the cultural ordering of society, and humans have created a great diversity of systems of time and space reckoning that evolved according to the precise contexts that shaped them. At the social level, control through the production of cultural systems of meaning and values also developed through the diversity of institutions, traditions and codes that humans have constructed. As societies became more complex, then more sophisticated forms of control became necessary. And so systems of government developed, as did forms of politics, law, institutions of control such as armies, prisons, hospitals, schools and so on, which became imbued with deep cultural significance that served (more or less effectively) to regulate and order societies.

In the period of modernity this growing complexity and interaction of cultural norms fed into what John Frow terms the general 'social organization of culture' (1995:1695). Arts and sciences were analytically (and culturally) separated, as was mass culture from high culture. Space and time became more fully developed *scientific* concepts. Newtonian-based science extracted these from the culturally diverse realms of which they once were part to become a rigid and universal mathematical set of principles. With regard to our relationships with time and space, this development resulted in the institutionalization of a formal geometry of space reckoning and the inculcation of clock time as the *real* measure of time.

Technology, 'man's mode of dealing with nature' as Marx put it, is of course integral to this process. Without the most basic technological aids there would be no cultural production which could be said to constitute forms of social control; even the metaphysical dimension of cultural control such as religion would be impossible without the physical construction of the symbols and signs that make them 'real'. And our relationship to technology and technological development has, for the most part in human history has been one that evolved through the dialectic of 'social shaping'. Increasingly complex and sophisticated technological 'extensions' as McLuhan termed them, allowed modernity to develop, bringing the control-oriented processes of cultural production into industrial processes. This development occurred early in the industrial revolution, but reached its apex in what could be called the 'culture of Fordism' that emerged in the post Second World War era. And as Harvey (1989:135) writes of this process:

> Post-war Fordism has to be seen...less as a mere system of mass production and more as a *total way of life*. Mass production meant standardiza-

> tion of product as well as mass consumption; and meant that a whole new aesthetic and a commodification of culture...

Note how Harvey implies a creeping imbalance between the dialectic of social shaping of technology as means of control through cultural forms—to a dynamic within high Fordism where technological forms begin to dominate, producing a 'total way of life' and 'a commodification' of culture. The webs we began to spin at this period were cultural webs, but increasingly they were being determined more than ever by the quest for economic and technological solutions. Fordism had thus been the means through which humans finally created what Neil Postman called 'Technopoly' or 'the submission of all forms of cultural life to the sovereignty and technique of technology' (1993:52).

By the early 1970s the system of Fordism was doomed. But the rise of technological forms of domination continued, of course, and indeed was massively compounded and expanded through the revolution in ICTs. The 'need for speed' through computer networks has rapidly created what Lash sees as an 'information society [that] has for its unintended consequences the information culture' (2002:146). The search for order and control through abstract and machinic solutions thus contribute to unintended consequences or, in the case of the network society, systemic blowback. Lash sums this up neatly when he observes that: 'Modernity is ordered: modernity's consequences disordered. The consequences of order are disorder' (p. 146). Put another way, one could say that in our neoliberal society the consequences of control are risk, hazard and chance.

Notwithstanding this persistent disorder in economy and society, the innate drive for control still underscores the neoliberal project. However, the recurring failure for neoliberal outcomes to match neoliberal promise is justified or legitimated, to borrow from Geertz, by using 'ideology as a cultural system' (1993:173–234). The ideology is based on a fundamental contradiction: effective control over the form and shape of an economy (and of society) will always be elusive if the ideology centres around abstract market forces—which serve to take control away from social, economic and political institutions. What has occurred in the period of globalization from the 1970s is that computer-based technology has filled the control void that the end of Fordism has left. Computerization consolidates to become a form of technology that we have become reliant on like no other. Ubiquitous computing now assumes a pivotal position in economic and political life, and consequently begins to dominate the forms of cultural production that reflect the dynamics of this technological dependency. This is cyberculture.

The creation of cybercultural practice now goes broad and deep. The iridescent signs, symbols and representations of cyberculture displace and marginalize other forms of cultural production. This is especially the case with people who are born into the network society and for whom the 'time-squeeze', dense interconnectivity, speed and rapidly developing forms of digital media are normalized and internalized aspects of life. The ideology of control actually functions well at this level, concealing the reality of growing powerlessness through relentlessly promoting a utopian discourse which engages with much of what people actually want—which is of course control over as much of their lives as possible. Being connected, being part of the network society, of cyberspace, is to be on the road to a bright future, to be in command—and all this feeds the illusion of control.

What is more, this is not an elaborate deception practiced by the business class. They either consciously believe it, or unthinkingly accept it as true. Whichever way, business can see that cyberculture accords perfectly with the project of neoliberalism. For them cyberculture translates as tech-savvy workers who are 'self-directed', autonomous, entrepreneurial, and able to make ICTs do all manner of wonderful things; to turn ICT applications into novel business practices, to develop new products and open up new markets and new demands for other businesses to service. These tech savvy network builders are what Louis Rossetto of *Wired* magazine called the 'heroes' of the information age, the connected denizens of the creative industries that governments put so much emphasis on and investment into. Theirs is a compelling worldview that reflects a growing 'Finlandization' of the planet, where ICTs are the principal drivers of the economy, transforming old business concerns (the Finnish company Nokia used to make paper and rubber products) into weightless business entities that span the globe and whose virtuality seeps into every pore and moves every process.

Let us look now an example of cyberculture in practice.

Serious games

In cyberculture a relentlessly grey homogeneity that emanates from instrumentalism lurks just beneath the colorful surface of diversity, opportunity, innovation and efficiency. Take game culture. This is a massively popular activity where players enter a virtual world linked through computers to compete in increasingly realistic games in genres such as motor racing and science fiction-based adventure. It is a prototypical cybercultural practice. Many of its practitioners are attracted to it because it has connotations with the lifestyle made famous in Douglas Copeland's anti-Baby Boomer tract *Generation X: Tales for an Accelerated Culture* (1991). Copeland inspires much of this cohort through the manifesto-like chapters in his book. Chapter titles such as 'I Am Not a Target Market', or 'Enter Hyperspace' or 'Purchased Experiences Don't Count' purportedly define the Generation X *zeitgeist*. Gaming is thus widely seen as a hip, non-mainstream activity where individuals can express themselves in true Generation X fashion and compete with each other using cutting-edge computer applications and highly trained hand-eye coordination skills in a context where a Marinetti-like glorification of speed pervades. Not for them, so the rhetoric goes, the slavery of conformity or complicity with the consumerism and commodification that had blighted the worldview of the baby-boomers.

Notwithstanding its 'youth' and its 'radical' connotations—mixed-up it should be added, with the fascination with some of the more lawless aspects of hacker culture—the real name of the cyber game is business. And it is an immense business. The games industry roundtable predicted worldwide growth (game consoles and software sales) to expand from $31.2 to $49 billion from 2006–2011. This would make gaming a bigger industry than Hollywood movies (Gallacher, 2008). The market moreover has enormous room for further expansion as the network society spreads and deepens. For example, the number of users in China went from zero in 2000 to 14 million in 2005 (Joseph, 2005). One 'free' game, when launched in China in 2007, attracted over a million registered users within two weeks (BNET, 2007). Moreover, industry leaders Microsoft, Nintendo and Sony, are deliberately seeking to expand the market out from its young male demographic (Joseph, 2005). As in any industry under capitalism, the tendency toward monopoly control always makes itself felt. It is an age-old pattern and an iron economic logic that emerges clearly in the gaming industry. The logic is as follows: as hardware and software become more complex and powerful, then the producers incur increasing development costs. The innovators of many of the popular games, however, are from small independent software producers comprised of less than ten people. These companies with possibly fresh idea have two choices: devote their entire workforce to a single project; or seek buyers for the business and be subsumed into a multinational. The first option is a considerable risk in a fickle industry, and the second is one that tends to iron out any kinks of creativity and originality in game design, and goes with what will predictably sell in the marketplace.

Users themselves are not immune to the effects or the dictates of business. The laws of globalizing capital flow through the networks of online game users along with the bits and bytes of information that drive the games themselves. Online gaming is the burgeoning centre of the games industry. Users are linked through the Internet and compete with each other. They play games where killing virtual opponents, finding secret keys, more powerful weapons and so on, make up much of the challenge. More powerful games allow players from anywhere around the world to join forces against others in these virtual global battles. Players coming from 'virtually' any point on the face of the Earth, can band together to form alliances in cyberspace. But we live in a world of uneven development and the structure of these virtual armies reflects and sustains the concrete reality of global class stratification and exploitation. Tony Thompson, writing in the London *Observer* in 2005 describes an entrenched development in game culture whereby wealthy users in developed countries can pay users in less developed countries (ostensible allies) to do much of the routine work for their characters (avatars). Thompson writes that:

> The most valuable commodity in [online games] is time, and this has spawned the rise of the virtual sweatshops. Every player starts with little 'virtual money' and few skills. Moving up to the next level involves carrying out dull, repetitive tasks such as killing thousands of virtual monsters. But thanks to companies such as Gamersloot.net, players now have an alternative. They simply pay someone else to do the dull work, and buy a ready-made character at a more advanced level.

In other words gamers in the USA or UK, for example, can pay gamers in Romania (the example in the article) the equivalent of 50 cents an hour to do the tedious work of killing virtual monsters for up to ten hours a day. Freed from the tedium of slaying enemies all the time, wealthy gamers can involve themselves in the higher levels of the game, where the virtual rewards are greater. The very business-like nature of online gaming culture is further underlined when we learn that avatars can be bought and sold on eBay for four-figure sums

(Winokur, 2003). The real-world stresses that accompany this virtual space have other—predictably human—effects. For example in June 2005 Xinhua News Agency reported that a Shanghai online gamer had been sentenced to death, with a two-year reprieve, for murdering another gamer who 'stole' his virtual sword, a 'Dragon Sabre' used to kill opponents. The 'thief' had already sold the weapon to another gamer for $870.00 (Xinhua, 2005). In Russia in 2008 there occurred a similar incident, when online duelers arranged to meet and fight in real life. One man traveled 'to Moscow from Ukraine to meet his rival [and] the confrontation ended with the Moscow man being beaten to death' (Russia Today, 2008).

Boosters for game culture and cyberculture more generally, are not restricted to the business world. For example Gerard Jones in his book *Killing Monsters* argues that the violence and the fantasy (the majority of online games content) should not be discouraged. This is because computer games 'give them the coping skills they desperately need' in the real world, to 'help...better navigate the world around them' (2002:101). Doubtlessly individuals develop certain skills in these virtual worlds—but what kind of skills, and what do they indicate to us about our real world and its cultural production? It is perhaps more reflective (and certainly less sanguine) to suggest that at some level, people realize that to survive in both virtual and actual reality (where the dividing line increasingly blurs) it is necessary to acquire the hand-eye coordination skill need to synchronize with high speed action, and develop the calculative (rat cunning) skills to outwit opponents in the hyper-competitive world of gaming. A sense of social solidarity is of course as absent from the virtual game world as it is from the actual world that neoliberalism constructs.

Gaming is about the survival of the avatar in an always hostile or dangerous environment, and likewise in the real-life world of individualism and acquisitiveness, we are schooled within network culture to seize the moment or risk being left behind. Successful gamers think and act quickly because in the real world this is what is required for their survival: both at the level of the individual and at the level of the business enterprise. Quick-wittedness is a fundamental skill learnt online and it is something now highly valued in the real world. This is a recent development. The personality trait of fast thinking, as Nicholas Lehmann (1998) observed, was once seen as suspect:

> It used to be that the business world was neutral, or even hostile, toward this quality [of quick-wittedness]. Especially at big corporations, quickness marked someone as peculiar, high-strung, and unreliable. But now the situation has changed. In high tech and finance at the very least, quick-wittedness rules [and] is not a state of affairs that would, in most times and places in history, have been considered normal and healthy.

But today it is. Major corporations such as Microsoft have developed questionnaires that are designed specifically to highlight this quality in prospective employees. Being able to 'think quickly' has its obvious advantages in sectors such as high technology and finance, as Lehmann notes. But in the accelerated network society where 'seizing the moment' is everything, such thinking is now the *sine qua non* in almost all sectors of the new economy. Boiled down to its essence this abbreviated-thinking-in-action is very near to a form of gambling or risk-taking and is what successful gamers do best. Computer games require these skills. They are developed on narratives that reflect the most speed-worshipping and violent aspects of our culture (car racing, gun play, space-age fantasy and so on). These narratives and the temporalities that propel them thus become second nature to young people brought up in the virtual environments of Playstation, Xbox, Wii or GameCube. Peer pressures and the addictive nature of the games themselves mean that users must quickly 'get up to speed' both cognitively and physiologically to compete and endure in this virtual combat zone, a zone that increasingly overlaps with reality.

Game culture is discussed as an element of cyberculture because it is seen by many as *something different* from the instrumental reality of day-to-day jobs in the network society. What is argued here is that cyberculture is stamped with the same logic of profit and loss as much as any other realm in the network society. Generation X and the cohorts beyond are consumers who enthusiastically buy into this logic. For its part, the virtual world of gaming is a component of the parallel world of business that impinges on our lives. Through its growing preponderance, it seals off access to other worlds and other ways of thinking about our own lives. The screen becomes the closed world of the quick thinking gamer, the stressed office worker, harried student or pressured executive alike. These connected worlds are the context for abbreviated thinking. We become what Madeline Bunting (2005) calls 'willing slaves' to a high-tech capitalist system that seems ostensibly to be democratic, full of choice and progressive, but which in fact masks a reality of anxiety and worry; a world of evanescence and depthlessness where the deeper currents of change are blurred by increasing speed.

The journalist Bunting calls this 'cult of efficiency' a 'massive fraud' (2005) whereby big business entices people into the high-speed network society with the seductions of consumerism: a sleek-looking iMac for you to become more productive and well-organized, and a concerned offer of professional counseling when you become burnt out or are made redundant. This form of coercion supposedly replaces the relatively more ponderous carrot-and-stick approach that had characterized worker-capitalist relations in an earlier phase. However, there is no organized deception on the part of big business or anybody else. No one is in charge. The system of feedback has broken down (or was never really established). The forms of human control that the information technology revolution was supposed to bring were never able take root because social institutions were consciously excluded from the development of computerization in the quest for open-ended speed. Globalization and the network society it has engendered, is out of control. Consequently, the roles of human 'steersman' and 'pilot' envisaged in their cybernetic dreams by Wiener and Licklider are being systematically abolished within a neoliberal economy. Their tasks become fully automated, and the steersmen and pilots experience 'restructuring' or are relegated to pulling the oars inside the hull of the ship. The problem is that, like slaves in the bowels of a galley, they must keep their heads down and pull ever faster under threat of being thrown to the stormy seas. Pre-occupation with their toil means they have no time to lift their heads, but if they did they would realize that there is no one any longer at the wheel.

We are dependent on computing to an extent that has no parallel in our human relationship with technology, and computing is the very lifeblood of the networked society. Barring a nuclear or environmental catastrophe it almost impossible to think of returning to a form of organization that is not built on its logic. Computers and high-speed network systems are now an ineffaceable fact of life, yet we have no large-scale or meaningful social control over their development. Where do alternative visions come from? In the eye of the storm that is the 'cult of information' where do we find the times and spaces to dream, to reflect, and develop critiques of this world and construct imaginaries of other possible worlds? As Scott Lash observes in his book *Critique of Information* (2002:1):

> The global information order has erased and swallowed up into itself all transcendentals. There is no outside space anymore for critical reflection. And there is just as little time. There is no escaping from the information order, thus the critique of information will have to come from inside the information itself.

The penultimate section of this book will look at the how the neoliberal 'information order' is being challenged (and critiqued) by millions of people around the world. The digital Empire of Speed is, for all its dearth of social control, a global *political* space. Indeed it is intensely political. Within it people are articulating critiques of neoliberalism and its specific market-oriented form of globalization (Klein, 2000; 2007). Importantly, the critique is developing through use of the information technologies and the networks that neoliberalism has created—in other words, from 'inside the information [order] itself'. It comes from politically minded tech-savvy users of networked computers and networked applications and practices. Their actions constitute a profoundly political challenge in an age where institutionalized politics, the politics of liberal democracy, has undergone deep crises. This is, then, is a new form of political activism. The reactions to the depredations of the networked order are taking politics to a new realm and a new plane of social articulation. These spaces are virtual and physical, local and global at the same time. It has different aims and uses different processes from earlier political forms. It is an evolving politics that has grown out of the crises of liberal democracy that emerged and evolved in an era that began in the 18th century, and therefore reflected a technologically and politically different world. This was a world that was already coming to an end by the 1970s. With the rise of neoliberalism, 'old' liberalism, or more precisely *liberal democracy*, has been found wanting by millions all over the world as markets and speed-driven networks ride roughshod over the forms of democratic control that had constituted our once-cherished and vibrant political institutions.

Something has happened to the Enlightenment project of liberal democracy; an interest and belief in what happens between our elected representatives in parliaments and congresses and senates has sunk precipitously. The alternative political space of the Second Empire of Speed is a nascent but energetic space where different ways and means to 18th century political practices are being sought (Saul, 2005). To understand why liberal democracy has come to such a point of global crises, and to understand the logic behind the countercurrents to neoliberalism, we need to look at the structures of liberal democracy itself. This is done once more from a temporal perspective. Accordingly, what the following chapter tries to articulate is that, just as time is in everything (*eigenzeiten*), in people, in nature and in the environments that surrounds us, time (or temporality) is in the political structures we create. It was argued in chapter one that to understand the nature of social acceleration, it is important to grasp the idea that technologies are 'time-loaded'. We saw that with the dawn of industrialization and modernity, the technologies that made them possible were 'time-loaded' by the dominating time of the period, the time of the clock. Liberal democracy grew and flourished alongside industrialization and modernity. As the following section shows, liberal democracy too has been 'time-loaded' in large part by the clock. It follows further that it is a form of political action that is nowadays literally too slow to function as a truly representative form of democracy in the 21st century, the century that is set to be dominated increasingly by networks, by market forces and by open-ended speed.

CHAPTER SIX

THE SPEED OF LIBERAL DEMOCRACY

Time is to politics what space is to geometry.
Regis Debray, (1973:90).

...political time is out of synch with the temporalities, rhythms, and pace governing economy and culture.
Sheldon Wolin, (1997)

Précis

As Debray's quotation indicates, time is central to the political process, and in the quote from Wolin we see the observation of a temporal disjuncture between politics and the economy. In this chapter it is argued in some detail why this is the case. It is suggested, moreover, that some disturbing consequences flow from this idea. To do this the temporal perspective is applied in particular to the processes of liberal democracy. The emerging theory attempts to make more explicit the rhythms and tempo—the contextually derived timescape, in other words—of liberal democracy, the preeminent political system in the world today. Liberal democracy is a form of political representation that came into being over the course of the 18th and 19th centuries. And like the dynamics of modernity and industrialism that have become so much part of its evolution, liberal democracy is 'time loaded' and 'conditioned' by the context of its creation. The timescapes that are at its core reflect a world that had a very different temporality from our own. As it evolved and spread, liberal democracy developed specific institutions and traditions that broadly reflected these bass-line rhythms. It was a form of politics that was able to lead and shape the economy, culture and society for much of the 18th and 19th centuries. Given the relatively unsophisticated technological levels of 'connectedness' that governed the potential of the temporal context at that time, liberal democracy could function 'fast' enough to develop this leading role. However, as we will see, over the passage of time, the institutions of liberal democracy are also prone to inertia. A major consequence is that as its institutions of power become slower in comparison with an always growing and accelerating industrialization process, then liberal democracy becomes less able to lead and begins to *react* more to the developments (and imperatives) of capitalism and capitalist modernity.

Through this perspective, the chapter reveals that during the post-World War Two phase, the era of 'high Fordism', there was something of an interregnum in the inexorable temporal disconnection of the polity from economy and society. Indeed during this period politics was able, albeit in cooperation with capital and labour, to *lead* once more and utilize the processes of what was a *social democracy* to make positive changes in the lives of people. Importantly, these decades constituted the zenith of the democratic process within the 'arc of time'—as Debray put it (2007:6)—that was most conducive to it delivering upon its historical responsibilities. The rise of neoliberal globalization, the ICT revolution and the network society during the 1980s signaled the end of the social democratic experiment. Social and economic acceleration has made the temporal disconnection even more profound and have left the institutions of liberal democracy trailing in their wake. The major consequence has been an increasingly undemocratic society where the historical responsibilities of democracy (to articulate the needs and wants of people) are abandoned in favour of a neoliberal 'democracy' that is concerned first and foremost with creating the right business conditions for capitalism.

No time for politics

If we consider once more the concept of timescapes as a method of perceiving the temporal dynamics of human action—individual and social—and if we apply this to political theory and to political history, then a couple of profound (and alarming) conclusions force themselves onto our understanding the nature of politics and democratic political agency today.

The first is that the politics of liberal democracy (the system of representative democracy that is held commonly to be the most advanced means devised of achieving justice and fairness for the majority) is becoming disconnected from the dynamics of economy and society in the age of globalization and the ICT revolution. Second, and related, is that liberal democracy is increasingly unable to fulfill its most basic function, which is to represent the civil and political rights of the people, the *demos*.

The temporal perspective provides this insight and shows that the increasing speed of life, of culture, economy and society is a central factor in this breakdown of democracy. Indeed this failure is also the failure for liberal democracy to guarantee the *temporal rights* of individuals and cultures and societies today. Time as a resource, time as a *human right*, a right that should be considered in the same way as the right to freedom of speech or the protection against slavery, is being systematically denied within an over-powerful and increasingly autonomous economic system, an empire that is predicated on the unrestricted application and dissemination of speed and profit throughout the social space of the planet.

The insight is surprising and then again, not. Herbert Reid, as we saw, noted the lacunae at the centre of our understanding between politics and temporality in 1973, prior to our current and much more pressing context. More recently Sheldon Wolin made a broad statement of the problem in his 1997 essay 'What is Time?' when he observed that 'the instability of political time' may best be approached 'through the language of temporality'. And Douglass North (1999) also tried to put the issue of temporality as a method for framing research and analysis of political processes back on the agenda. He sees our present situation as a bleak one:

> For an economic historian, time has always been something that is fundamentally disturbing, because there is no time in neoclassical theory. The neoclassical model is a model of an instant of time, and it does not therefore take into account what time does...I will be blunt: Without a deep understanding of time, you will be lousy political scientists, because time is the dimension in which ideas and institutions and beliefs evolve. (cited in Pierson, 2004:1)

One could take issue with the last sentence in this quote, because as the logic of this argument so far would suggest, 'ideas and institutions and beliefs' do not evolve *in time*—but that time evolves (is in the process of becoming) *with* 'ideas institutions and beliefs'. But this may be to quibble with the general thrust. North makes his point rather well, and identifies a blind spot in the ways that social science analyses history, especially as it relates to the theory and practice of politics. Here the central argument is that politics—as at once an 'idea' an 'institution' and a 'belief', in the dominating model that has come to be known as liberal democracy—has indeed evolved with the 'passing of time' (since the late 18th century) but has done so only glacially and through force of pressure from below.

That politics has a specific temporality is the central question at stake here, and in many ways is the hinge issue of this book. We are still finding our steps here. As Reid, Wolin, North and others have indicated, the approach begins first with the identification the problem—which is the nexus between politics and economy within the frame of temporality. The next step, and is the step that much of this book has tried to take, is to fashion a theoretical framework that helps explain the problematic. This (no pun intended) takes time, and it requires us to go back to basic principles that give shape to the theory and the argument. Consequently, wherever possible I will link that temporal framework to concrete situations so as to illustrate more clearly what we are dealing with conceptually—and judge how well it is able to explain empirical reality.

As we shall see, neoclassical politics, in the shape of liberal democracy has inherent tendencies towards *inertia* and *crystallization*, a propensity to try to hold back the 'flow' of time and change in ways that reflect its own unique context of becoming. In other words, a mutually implicating acceleration of culture, technology, economy and society has pulled clear from the fossilizing mechanics of classical political organization, ideas and agency that were developed in another time and at another speed. To understand how this occurred it is necessary once again to become 'time aware', and to look at the issue of politics with the theory of temporality, of timescapes and of *eigenzeiten* in the foreground of the analysis. But before we do that, it may be useful to look at the term 'liberal democracy' and make it clear what is meant by the term before subjecting it to temporal analysis.

What is Liberal Democracy?

There is no simple definition of the term, and much debate continues that reflects the complexities of what liberal democracy is and what it does (or should do)(Lipset, 1960; Bobbio, 1984; Habermas, 1987; Held, 1987). These need not detain us here. What can be said is that in general terms, for many people in the Westernized developed and developing economies of the world today, liberal democracy is seen as something akin to the highest form of political organization that humans can aspire to. It is seen as the benchmark of a civilized country. States that do not have liberal democratic systems are viewed negatively as 'authoritarian', or as 'dictatorships' and are regularly sanctioned for having (by default) a heretical political system; or, as in the case of several states of sub-Saharan Africa, they are labeled as 'failed' states whose shortcomings evoke the patronizing 'conditional' disbursement of aid with which to ease (or, as some argue, prolong) their suffering.

The terms 'liberal' and 'democratic' can be separated out if we look at the contemporary political scene. Let us look first at 'democratic'. Aspirant developing countries must show their 'democratic credentials' (at least *de jure*) if they are to have normal diplomatic relations and receive multilateral aid from bodies such as the World Bank and the International Monetary Fund. Core elements of democracy such as representative government, a universal franchise, regular elections, an independent media, relative transparency in governmental processes and so on, are seen as the minimum standards required to be called democratic. Of course the practice can differ from the theory, and states such as China or Russia can practice repression and other undemocratic measures, whilst ostensibly having democratic constitutions. Hypocrisy can also be involved in such country-to-county relations, and in many instances democratic principles are put aside should they not fit with more pressing political and economic interests. A case in point here is that of the monarchy of Saudi Arabia where an overtly (and constitutionally) undemocratic country receives normal, if not sometimes favorable treatment from the majority of the world's liberal democracies.

On the other hand, to qualify as a 'liberal' democracy, to be seen as the kind of country and polity that others should aspire to, then the preconditions are self-set. And so not surprisingly the leading liberal democracies are those who have the power, the prestige and the history to identify themselves as such. Moreover, such polities agree on who else is to be part of this inner circle of political and cultural sophisticates. Countries and regions such as the USA, Canada, Britain, Australia and most of Europe, by general consensus amongst themselves, are democratic polities; and, by implication these are said to contain the most highly developed and complex forms of civil society. Moreover, the term 'Liberal' comes freighted today with notions of free trade, open markets and individual freedom, as opposed to statism, managed economies and collectivism.

As I noted briefly in the Introduction, Francis Fukuyama conspicuously enunciated the ideology of liberal democracy as being the fullest expression of human political potential in his 1992 book *The End of History*. In it he asserted that history in the last quarter of the twentieth century had in fact come to an end due to the triumph of liberal democracy in the wake of the imploding socialist experiment. Liberal democracy, he maintained, constituted the 'end point of mankind's ideological evolution' and the 'final form of human government'. Indeed, the '*ideal* of liberal democracy', Fukuyama insisted, 'could not be improved on' (1992) (italics in original). At the time his thesis was extraordinarily controversial and divisive in academic, policy and political circles. This is because on one hand for some people it merely expressed a set of 'truths' that, for example, America's political founding fathers saw to be 'self-evident'. Alternatively, others saw it in terms of a frighteningly existentialist scenario, engendering a sort of 'there must be more to life than this' attitude that required taking Fukuyama's thesis to task, primarily through various Marxist critiques and historiographies (e.g. Callinicos, 1991; Derrida, 1993). Fukuyama's thesis broke open an ideological hornet's nest mainly in the Western academy, in policy circles and in think tanks—each using it or abusing it to justify their own particular positions regarding the value of liberal democracy, as they perceived it. Two decades later, however, the buzz has died down, and in our nownetworked society, the 'self-evident' truths of liberal democracy, the stock phrases of politicians mainly, are *still* purported and broadened out as the only valid political direction for the *demos* and for civil society. The pity (and the problem) is that political life and civil society in general has never been so fractured and divisive, so full of uncertainty and inequality—notwithstanding all the formal protections that liberal democracy is supposed to afford (Sennett, 1998; Putnam, 2000; Saul, 2005).

All this then begs the question: why does there seem to be such a gulf between the theory and the practice of liberal democracy? To unpack this contradiction it is necessary to step back a little bit to look at some of the major factors that went into the formation of liberal democracy in the 18th century. In particular it is necessary to look at the beginnings of liberal democracy through the prism of temporality. What were the primary timescapes that dominated the economy, culture and polity at this time, and what kind of contextual timescape did these construct? What kinds of speed dominated at this formative period and how were they generated? Lastly, it is necessary to set this timescape and its specific tendencies against the changing social, economic and cultural forces of capitalism that were born alongside classical liberal democracy, but were driven by their own internal logics.

The roots of classical liberal democracy

Fukuyama's thesis has its roots in a faith in classical Enlightenment thought. This promoted the idea that the perfectibility of human systems was basically a matter of uncovering their immutable 'truths' that had been hidden by mythology, ignorance and religion. The 18th and 19th century was a special period of history that saw the confluence of a whole range of revolutions in the ways in which people oriented themselves to the world. Under the general rubric of the Enlightenment, ideas in science and technology, philosophy and political economy, secularism and political theory, triggered an intellectual process that would transform human society. It was possibly the most significant time in world history in terms of its long-term legacy. It was a self-conscious transformation of what it meant to be human in a proto-modern world. Indeed, whereas historical periods such as the Renaissance or the Middle Ages were designated *post facto* by historians, the Enlightenment was possibly the only time in history where the leading figures themselves actually designated the time in which they were living through as one of 'enlightenment'.

The somewhat self-aggrandizing *zeitgeist* that pervaded the intellectual climate of the Enlightenment was one of boundless optimism and seemingly unlimited possibilities. The human mind (through the application of logic, reason and critical thought) was deemed able to solve any problem and deliver humankind from charlatanism and superstition. For example, it was Voltaire, a leading French author and philosopher of the Enlightenment, who encapsulated the ethos when he wrote that: 'No problem can withstand the sustained assault of thinking', encapsulated the ethos. The Enlightenment, moreover, was totalizing in its intended field of effect, expressing as Adorno and Horkheimer put it, 'the actual movement of civil society as a whole in the aspect of its idea embodied in individuals and institutions…' (1986:xiv).

Science, commerce, art, literature, philosophy and politics were all affected, and these in their turn would revolutionize culture and society. This was an intellectually effervescent period when the *philosophes* first began to think about (or to revisit) ideas such as mass education, penal reform, secularism, republicanism, nationalism, freedom, tolerance, scientific method and constitutionalism to name but a few. It was an age where humanity seemed to wake up and view the world in a new way, in the clear light of reason and commonsense. Society and its institutions, it was increasingly assumed, could be remade for the better; they could be newly-invented from scratch to fit with the novel circumstances—or be destroyed altogether if they did not accord with the new thinking.

The institutions of politics, which had traditionally served the interests of an elite stratum, were now a core domain for such reforming and reinvention. Politics was the power source for much of society. As a social process, politics was able to marshal the material resources and the ideological will that could implement the changes that would transmit the vibrant pulse of the Enlightenment to many other fields of life. What this meant in practice was that political power concentrated in the hands of the few was giving way to ideas for an emancipatory politics for the benefit of the majority.

A specific view regarding what time was, lay at the heart of this revolution in thought. Time in Enlightenment thinking was energetic, and it flowed from the universal and mechanical principles of Newtonian Physics. Time *unfolded* leaving history in its wake; it oriented its flow towards a luminous future that beckoned constantly on the horizon. This made for a dynamic present. In Enlightenment thinking in general and in political thought in particular, this could best be described as being an 'open' present. Accordingly, political philosophers would freely reach into the past for guidance on how to reshape their present and then direct it towards a knowable and plannable future.

Classical Greece and Rome served as exemplars for principles of government that would act as the basis of neoclassical politics. For example Greek conceptions of democracy, of *demos*, of government by the people, were appropriated and refashioned thinkers such as John Locke and Montesquieu, as meaning government through *consent* by the people. Montesquieu in particular looked to Britain after the English Glorious Revolution of 1688 as the model for modern politics and governance. In his *Spirit of the Laws* of 1747 Montesquieu praised the British constitution that separated state power into three independent branches of government: the executive, the legislative, and the judicial. The fact that no single person or cabal was in charge, he reasoned, established a system of 'checks and balances' that gave the maximum amount of freedom to the general population. It was the very absence of such a constitution in France at this time, together with the popular perception of a corrupt, decadent and absolutist monarchy, which made that country the locus of much radical Enlightenment thought.

This was an ever-yawning void between the people and the elites that culminated in the Revolution of 1789.

The Roman Cicero's ideas on civil society were revived and redeveloped by thinkers such as Thomas Paine, G.W.F. Hegel and Alexis de Tocqueville. Civil society came to be seen as an important domain that existed *outside* the forms of state and governance. Indeed it was characterized by a separation between the state and society that retains strong resonances today. David Held (1987:281) describes it in more detail:

> Civil society retains a distinctive character to the extent that it is made up of areas of social life—the domestic world, the economic sphere, cultural activities and political interaction—which are organized by private or voluntary arrangements between individuals and groups outside the direct control of the state.

During the Enlightenment period, the 'economic sphere' became, rapidly, the most dynamic and powerful element of civil society. Indeed, the growing power of civil society, beginning in the 18th century, conferred a hitherto unknown level of freedom and rights to industry and commerce. The Enlightenment and the modernity it brought into being was undoubtedly a world-changing phase in human history whose effects have resonated down the centuries to our own time. It is also a period and a legacy that has been analyzed extensively from its political, technological, economic, philosophic and industrial perspectives. What have not been analyzed to anything like the same extent are its temporal dimensions—the nature and quality of the timescapes that developed from this convergence of social forces.

The speed of classic liberal democracy

The merged dynamics of Enlightenment, industrialization and modernity had the restive logic of acceleration at its core. Instrumental efficiency, secularity, clarity, organization, transportation, planning and progress were just some of the watchwords that informed its bubbling and surging spirit. Politics, or the marshalling of society's powers to realize society's newfound potentials, drove and was driven by these dynamics of modernity, speed and industry.

What were its constitutive 'times' during the period, say, from the late 18th century to the early 19th? A few sentences are needed first of all to outline the temporal analysis that will illuminate this particular problem. To begin with, it is necessary to understand the importance of social *connectedness* during this phase of history, a period that saw the term 'net-work' as fairly recent coinage (the mid-16th century). The actual density of social connectivity is dependent upon technological innovation. In other words, the greater the sophistication of communication technologies becomes, then the greater the levels (and density) of communication in the creation of a network ecology. Second, this ecology, at whatever level of sophistication, comprises a timescape, wherein a context is created where interpenetrations of temporal relations 'cluster' into specific temporal relationships that have their own contextually shaped and shared rhythms. These can be powerful, dominating and long-lasting rhythms such as that of the clock, or they can, through the nature of the context which causes the timescapes to merge, be much more evanescent and short-lived.

In the period we are concerned with here—the century that encompassed the beginnings of the industrial revolution, through to its golden age of steam—socially constructed space and time were undergoing their own intense periods of technologically driven revolutions. To begin with, in terms of connectedness, in the late-18th century when Montesquieu was cogitating on his *Spirit of the Laws*, the concept of *globality*, of a 'consciousness of the world as a whole' as Roland Robertson (1992:8) put it was still a very rarefied one. A thin social stratum of intellectuals, philosophers, navigators, traders and explorers would have been aware of the world as a planetary whole. Henry of Mainz, for example, had produced the first map of the world in 1110 AD, and new and more accurate chartings had appeared regularly ever since, reflecting the growing (but relatively slow) process of diverse human societies coming into contact and connecting with each other (Livingstone and Withers, 1999).

However, for most people, including intellectuals and philosophers, the idea of a world where everyone would be connected, would have been very much an abstract one. The 'world' in practice for the vast majority of people was still very much a localized one.

From this Eurocentric perspective, the levels of connectedness between villages, towns, cities, regions and nations were relatively loose—and in many cases non-existent. Australia was not 'discovered' and annexed until 1788, for example; and at this time much of Africa and Asia was still to experience the unasked-for contact of the European mercenary, missionary and merchant. Systematic networks of communication flows and information exchange, not to mention commodity production, consumption and exchange, were still fairly rudimentary. The central point is that these connections were both tenuous and capacious; they contained the potential spaces and times with which to accommodate human actions at a relatively slow tempo that was suited to, and reflective of, the existing political and technological contexts.

The temporal *flow* of information and networks of communication becomes salient here because in the development of liberal democracy, the currency was *ideas*. This complex notion has been expressed clearly by Regis Debray who writes that it is:

> Impossible to grasp the nature of conscious collective life in any epoch without an understanding of the material forms and processes through which its ideas were transmitted—the communication networks that enable thought to have social existence (2007:9).

In Debray's view technologically generated speed (temporal rhythms) has a powerful role in the forms of ideas we conceive of and how they are given social and material reality. And in a historical context dominated by writing and the printed word, a phase Debray calls the 'graphosphere'—that great 'arc of time' that began with the invention of moveable type in the 16th century, until the arrival of the electronic 'videosphere' in the late 20th—specific clock-centered rhythms constituted the 'material forms and processes' that saw the emergence of liberal democracy (Debray, 2007:5–6).

What were the salient temporalities of this period of the 'graphoshpere'? Well, information and the ideas that it carried was, for widespread distribution, necessarily print bound. Accordingly the ideas of democracy were conceived, fixed in print, and given life through practice in the social world. As Debray puts it, the beginnings of the graphosphere laid down the basis for 'the age of reason and of the book, of the newspaper and *political party*' (p. 6). (Italics mine) How were print-bound ideas of democracy (or anything else) disseminated through the increasingly complex social networks that were emerging at dawn of liberal democracy? The first thing to realize is that a *relative slowness* permeates this process, and this was expressed in the embedded temporalities that the theory and practice of liberal democracy maintained. For example, in the mid-to late-18th century, information was communicated through means that had been available for a long time. For example, people walked on foot to meet with others to discuss the ideas and issues of the day. Horseback, for those who could afford to use it, took people longer distances within shorter timeframes. Sailing ships took influential political thinkers such as de Tocqueville and Paine from Europe to the nascent American Republic to spread their ideas and to imbibe new ones in the places they visited and with people they met. However a ship traversing the 5000 miles from Southampton to Boston could take a very uncomfortable six to eight weeks. Once in Boston, the variable state of the roads and the time of the year could affect a trip to, say, New York. This was a 250-mile trudge that could take another week or more (Isserman, 2002).

And in the 'graphosphere' the vectors for ideas existed in much more potent form in books, treatises and pamphlets that spread the tenets of liberal democracy to a much wider audience than any individual could. These too were dependent on the speed-generating technologies of mail coach, ship, booted feet and shoed horse. Nevertheless, in the context of the time, these ideas were traveling rapidly. For example, the news that George Washington's armies had defeated the British General Cornwallis's troops at Yorktown on 19th October 1781, reached King George III nearly six weeks later. If judged by our own standards of speed, the time gap would have rendered Washington's victory for nascent liberal democracy ancient history, or very old news, something that would have much-diminished relevance. But for those concerned at the time the news would have been fresh and something that could still be acted on; it would still be 'live' in that it would have teleological effects in a present that still connected deeply to the past and to the future in ways that hyperspeed tends to dissolve.

The point is that the ideas of liberal democracy—from the elite *philosophes* that generated them, to the foot soldiers of General Washington, or the *sans-culottes* of Paris in the French Revolution of 1789—could be discussed, read, disputed and *then* acted on. This was done within a certain context-created timescape that was at least amenable to a mode of reflection, to a working through of ideas towards a fuller understanding them and more systematic dissemination of them. The timescape both contextualized the ideas and the ideas themselves reflected the tempo of the timescape. Of course this does not mean to say that the political process did not make mistakes, or misjudge events, or act precipitously. However, what this does mean is that the time to reflect on the consequences of certain actions existed in a way that has been inexorably diminishing through the infusion of increasing speed throughout the project of modernity. In its elemental forming, then, liberal democracy developed an inherent temporality that reflected the technical, political, philosophical context of the age. If the institutions were social instead of supra-political, then the 'superceding' of its timescape may not be of great consequence. For example, a debating society such as the 18th century Scottish 'primitivists', who idealized the concept of the 'savage' and his supposed more noble traits, could come and the go when the social and intellectual context changed—it was of no great consequence in political terms. However, democracy was supposed to be timeless, and be somehow transcendent of mere social context. It is this contradiction that we will continue to explore in this chapter.

The slow-beating heart of liberal democracy

Systems of thinking such as logic, rationality and dialectic, the essences of the fresh currents that were transforming the world at the time of the Enlightenment, were derived from ancient Greece, from Plato, Socrates and Aristotle. These, too, of course, had their own temporal traces, their own times that evolved through their development as formal thought processes that sought to uncover the reality and truth of the sensible world. For example, a formal system of logic seeks empirical proof and validity (and reflection on the principles of validity itself) (Neale, 1984). The Enlightenment has often been referred to as the 'Age of Reason'. Deductive reasoning is a form of logical thinking, a method by which to arrive, as David Hume (1711–1776) put it '... at the discovery of truth or falsehood' (Gare, 1996:138). And similarly, dialectical thinking, a system developed by Socrates, seeks truth and reality through dialogue, a back-and-forth discussion comprised of proposition and counter-proposition, whereupon the 'truth' eventually emerges through the drawing out of the intrinsic contradictions in the other's position.

These are gross simplifications, of course, and each of these 'essences' has their own immense traditions and different branches of development that reflect both similarities and incompatible tendencies. The objective, however, has been to show that these all have common qualities. One such trait is *reflection*, the application of the mind to careful consideration of the complexities of a problem; the functional necessity to think though the issues at hand in the search for contradictions, inconsistencies and possible logical consequences. Another common characteristic is that these traits contain *no inherent speed dynamic*, no pressure or tendency, latent or active, which would automatically valorize rapidity or instrumental efficiency. They are intrinsically resistant, in other words, to the pull of speed. In respect of political time, Wolin argues that:

> Political time...requires an element of leisure, not in the sense of a leisure class...but in the sense of a leisurely pace. This is owing to the needs of political action to be preceded by deliberation and deliberation...takes time because it occurs in a setting of competing or conflicting but legitimate considerations (1997:4)

Classical liberal democracy has its own time, then, and represents, as North puts it 'a model of an instant of time' (2004:1). And as Fukuyama avers, it is an instant of time that exists as an *ideal* that 'could not be improved on' (1992:xi). Time is thus required to stand still with the theory, the practice, and the ideal. Indeed, these essences of liberal democracy could be said to be 'timeless' in that the 'time taken' to arrive at a 'truth' or 'fact' through logic, reason or dialectic or debate, is *immeasurable* because each issue, each set of contexts that shape each problem or question has its own time and its own unique temporal make-up. The underlying assumption concerning the application of logic, reason, or dialectic is that the 'time taken' is the 'time needed'—however long is necessary to eventually and properly come to the 'explanation' or 'proof' or 'reality' of a particular question.

Within its deep structures, classical liberal democracy has some identifiably 'embedded temporalities'. William Scheuerman in his *Liberal Democracy and the Social Acceleration of Time* (2004:49) makes this extremely illuminative point with reference to the so-called 'separation of powers', the single most vital element of liberal democratic functioning. The branches of liberal democratic systems of government that comprise the judiciary, the legislative and the executive, Scheuerman argues, all have specific and separate temporal orientations. The fact that the judiciary, for example, is oriented towards the past, is exemplified in its 'association with retrospective decision-making' and its functional emphasis upon tradition and precedent. Legislatures, for their part, are future-oriented in that they are expected to plan, and project and anticipate (as much as is possible) the future consequences of the laws that they conceive, debate and enact. And the executive branch is geared towards the present, towards making rapid decisions and responding to unforeseen events or crises, and ready to 'swing into motion a moment's notice' (Scheuerman, 2004:54) in circumstances where there is judged to be no time for prolonged legal deliberation or lengthy legislative supervision.

And at this level, within the workings of the parliament, the temporal rhythms derived from its context and from its overarching structures, evolved in many different ways that are commonly overlooked, but which are themselves embedded temporal forms. We see examples in the central institutions that liberal democracy has developed in the form of a parliament, or an assembly or congress of people's representatives. To these fixed places people physically convene to acquire information, to debate, to table motions, produce Bills for 'readings' which would then entail more debate, more time taken to 'reflect', 'digest', 'mull over', 'consider' and argue the pros and cons in a practical working of the Socratic dialectic in the search for truth and the right path towards progress and concord. Procedures such as the 'moratorium' from the Latin 'to delay', or the American 'filibuster' or British 'guillotine' became practices for being deliberately obstructionist and anti-speed, to avoid rushing to judgment or make hasty decisions in their legislative deliberations. The reasoning behind this is clear. As Scheuerman notes:

> Slow-going deliberative legislatures, as well as normatively admirable visions of constitutionalism and the rule of law predicated on the quest to assure legal stability, mesh poorly with the imperatives of speed, whereas anti-liberal and anti-democratic trends benefit from it (2001:2).

These slow-beating and 'leisurely' characteristics at the heart of the Enlightenment and at the core of liberal democracy became, by comparison with an emerging liberal capitalism, an increasingly unresponsive inertia that foreshadowed inevitable contradictions and their ultimate disconnection. Nevertheless during the 18th and 19th centuries, these 'timeless' metaphysical methods for discovering the underlying reality of the world, and of the limitless potential for the improvement of the human condition, were still at the core of the project of liberal democracy.

Dromocracy and the age of modernity

The new elite of the age of Enlightenment was comprised of philosophers, social engineers and the growing ranks of what Eric Hobsbawm (1996:2) called the 'conquering bourgeoisie'. These were emerging as a whole new vibrant and motivated class, what Virilio termed a *dromocracy* (1986) who were transforming the world, creating in the process the first Empire of Speed. Open-ended speed was not at the heart of liberal democracy, but its revolutionary ideas nonetheless unleashed countless social, cultural, economic and industrial processes that would have this logic built into them, and would eventually leave the 'leisurely' pace of social democracy far behind. At the beginning, however, the print-bound and print-disseminated ideas and rhythms of liberal democracy would lead the way.

A politics infused with the spirit of liberalism would clear the weeds of sloth and irrationality to forge a path to the future. In this way it was able to shape the 'pace of events' as opposed to merely trying to react to them as institutionalized politics does today. In an 18th and 19th century world that was still loosely connected in terms of its information flows, networked markets and distribution chains, liberal democracy (notwithstanding its glacially changing internal structures, extending the franchise, holding more regular elections, and so on) was still able to be a dynamic and legitimate force for change in the world of commerce, culture and in society more generally. Indeed it would be true to say that political dromocracy and industrial dromocracy were functioning in something approaching synchrony during the period of early modernity—and this was to the speed of the clock.

> During the formative years of modernity, the temporality of the clock fitted neatly with the dromocratic projects of politics and industry, and its logic and its influence grew in tandem with the rise of modernity itself. However, the clock also served to entime the politics of liberal democracy and the liberal democratic state. In fact the adoption of Greenwich Meantime as the coordinating time-structure of the industrial economies of the late 19th century, merely formalized a clock time dependency that had been inserting itself into modernity for a hundred years prior (Whitrow, 1988:157–169).

Clock-time on a general social scale, spreading as the validation of Newtonian physics and the mechanistic representation of a harmonious universe, meant that the future was something that was plannable, something that was unfolding into a knowable state of affairs. During the industrial revolution the clock was the rhythm of the new and the meter of progress. It signaled, with its increasing accuracy, the idea of modernity in action and industry on the move. Modernity was pushing ever forward, its ethos claimed, with boundless energy and infinite enthusiasm in the search for what *homo politicus* and *homo economicus* could achieve in partnership under the shared ideological banner of liberal democracy.

Temporally speaking, politics was successful and the political dromocracy could flourish because the 'pace of events' could still be created within and contained by the meter of the clock. Functioning as the temporal base-line of a developing modernity, the clock ticked away in a precise and orderly fashion that allowed not only political organization to flourish, but for the rational, instrumental and organizational propensities of an industrializing economy to grow in complexity, scale and depth. This progressed to the point where, around mid-to late-19th century, the 'Age of Capital' as Hobsbawm has termed it, moved into its first full phase of development in Europe and North America with the improving of the factory system and the full utilization of steam power. The dual dromocratic forces of liberal democracy and capitalist industrialization became the dominant political and economic forces in the world. Capitalist nation states were born and consolidated, cities grew and prospered, and populations increased and were delivered into the predictable and absolute rhythms of the clock and industrialization and modernization. Citizens internalized this time and viewed it (through a liberal democratic philosophy, as well as everyday subjective experience) as 'natural'. It was a world where the past, present and future could be represented as one long chain of human progress. What Virilio called the 'dromocratic revolution', produced for the first time in history, a *systematic* manufacture of machine-based speed that began to assert itself—and insert itself—into the everyday lives of people in culture, economy and society, through uncountable different social processes and technological applications.

Social acceleration, political inertia and the maturation of modernity

The central tenets of liberal democracy, those of democracy, liberty, freedom of association and speech, and so on, are today very much unchanged from their 18th century origins. And these in turn drew from much more ancient theories that have their roots in Greek and Roman antiquity. On one hand many would find this unsurprising. After all, these are the purportedly immutable essences of human social and political organization that is (or should be seen as, timeless), the system that, as Fukuyama saw it, 'could not be improved on' (1992:xi).

On the other hand the world today would be fairly unrecognizable to a person living even fifty years ago, much less two hundred or two thousand. So why do we accept (or are continually encouraged to accept) that liberal democracy is still the appropriate social and political organizational principle for the present age? The issue becomes even more pressing when we consider that liberal democracy *palpably* has not fulfilled its promise of a happy and contented present and a bright future. Why, after two hundred years and more of political democracy do its promises of a better tomorrow ring hollow and cynically in the ears of increasing numbers of people, especially young people? Why do our individual and collective futures seem to contract onto a frighteningly close horizon, a future time we feel unable to shape or even sense that we will be welcome within? What went wrong, in other words, with the promises of the Enlightenment, modernity and modernization?

Hartmut Rosa has pondered similar questions and he observes that:

> ...contrary to the other constitutive features of the modernization process—individualization, rationalization (functional and structural) differentiation, and the instrumental domestication of nature—which have all been the object of extensive analysis, the concept of acceleration still lacks a clear and workable definition and a systematic sociological analysis. Within systematic theories of modernity or modernization, acceleration is virtually absent; with the notable exception of Paul Virilio's 'dromological' approach to history...we cannot adequately understand the nature and character of modernity and the logic of its structural and cultural development unless we add the temporal perspective to our analysis (2003:3–4).

In this book I have sought in a modest and initiatory way to apply the temporal perspective to issues such as these. We can develop this further by using 'social acceleration' (Rosa's term) as a point of reference with which to view increasing speed as characteristic of social, cultural, political, economic and technological processes under capitalism. This helps us gain insight into how classical liberal democracy fails in its historic responsibilities, and falls short in its efforts to be relevant to the contemporary age.

The speed of liberal politics does not have an open-ended spectrum along which to accelerate—unlike the speed of capitalism. Its nature (if it is true to its tenets) is reasoned, deliberative, reflective and leisured. Notwithstanding the purported partnership of *homo politicus* and *homo economicus* throughout the project of modernity, liberal democracy is fundamentally and irreconcilably antithetical to the speed of capitalism. The ever-contracting instant produced by social acceleration dehistoricizes and defuturizes the present—all of which have an important bearing on the functioning of the speed and temporal frame of democratic politics. Jean Chesneaux states the problem, accurately and starkly, as it exists in our own time:

> Speed has become one of the paramount values and requirements in our modern societies. Yet democracy needs time, as a major pre-condition for political debate and decision-making; it cannot surrender blindly to speed. Nor does speed favour the dialogue between present, past and future, which is fundamental for the proper exercise of democracy (2000:407).

This disconnect between polity and economy that Chesneaux views as the central political issue of today did not occur suddenly as a cataclysmic break; the economy did not suddenly cut loose from its liberal democratic foundations and hurtle off into another realm of speed at levels approaching instantaneity. If the rupture had been sudden and total we would surely have noticed it. The process has in fact been much more gradual and corrosive. Liberal democracy and the economic system of capitalism were in theory supposed to be co-equivalents—mutually reinforcing dynamics that would propel humanity to an anticipated future of prosperity and happiness. Accordingly, the fiction that capitalist countries were 'naturally' democratic and that the truly democratic countries were 'naturally' dominated by free market capitalism could still be maintained. Even today, the disconnection, where it becomes apparent, can be attributed, usually successfully, to simple bad government or corrupt CEOs. The underlying assumption is that the 'system works' and will work if more 'responsible' governments are elected and those few 'bad apples' are moved out of the corporate boardroom to the prison farm.

The cleave between the time-consuming propensity at the heart of liberal democracy and the uncontrolled and open-ended obsession for speed at the core of what has become today a *neo*-liberal dominated capitalist system is only recently becoming apparent (Harvey, 1989; Bauman, 1998; Chesneaux, 2000; Scheuerman, 2003; Hassan, 2003). The gap widens as we embark on the 21st century and the uncertain, unsettling and unknown futures that this time holds. We move now to discuss in more detail the intrinsic and conflicting temporal qualities between the politics of liberal democracy and what is now generally recognized as neoliberal capitalism.

The triumph of speed over reflection

Let us begin with what is argued to be the generative motive force in the contemporary estrangement between economy and polity: neoliberal capitalism. The term *neo*liberalism is of course of recent coinage, its provenance being in the very deep intellectual synthesis of liberal democracy and market capitalism that has just been elaborated upon. However, from the period of around the 1860s—the mid-point of Hobsbawm's 'Age of Capital'—'liberal capitalism' is the term used to describe what we see in retrospect as nothing less than the triumph of capitalism as the paramount mode of economic (as well as social and political) organization. In his 1944 book, *The Great Transformation*, Karl Polanyi revealed the myth of the liberal-capitalist state as being one devoted to classical liberal values. Instead, he insisted that during the Age of Capital democratic institutional interests were sacrificed to the overriding interests of business. Such freedom meant, in part, autonomy for capitalism to spatially expand. Inevitably, this freedom was transformed and organized (by the state) as imperialism, and as Tony Judt observes '...it is hard to be an imperial democracy' (2004). Thus the growing economic power of the industrial capitalist class engaged in the day-to-day fight for markets and profits slowly but surely began to eradicate the traditions and legacies of the humanist, liberal Enlightenment. Indeed, Polanyi noted that: 'To narrow the sphere of the genus *economic* specifically to market phenomena [the workings of capitalism] is to eliminate the greatest part of man's history from the scene' (Polanyi, 1957:6).

This has been a deeply temporal shift as it served to sacrifice the past and the future on the altar of acceleration and instrumentalization. A principal effect, as Polanyi implies, is that dogmatized economic exigency is raised to the status of both means and end; the 'how' question predominates, and this is driven by the logic of speed. Accordingly, the facility or inclination to ask the 'why' questions become increasingly difficult, as these are considered marginal, if not irrelevant to the main game of life on Earth, which under the rule of capitalism is to increase material wealth. And as societies accelerate, the present becomes overwhelmingly where our focus lies. Ironically that same dearth of reflexivity due to the effects of speed makes the *ideology* of liberal democracy much more successful and the myths of progress, freedom, an open and exciting future, and so on, much easier to propagate and perpetuate. Indeed we come to believe in the ideal of classical liberal democracy *more*, even as our direct and meaningful experience of it dissolves in the rush of events that propel us through our lives. This was something that Terence K. Hopkins and Immanuel Wallerstein noted when they reviewed the broad historical context of this ideological progress:

> Liberalism's success as a mode of containing discontent and tumult was a direct function of the evidence its protagonists could present of incremental social 'progress', a thesis they were able to support quite well in the nineteenth century, but which has been much more difficult to argue consistently in the twentieth century (1996:8).

An important consequence of this ideological process was that time was perceived to be an important *economic resource* that had to be used efficiently and oriented toward material ends. This constituted an intensely utilitarian approach to time that, as noted previously, was famously analyzed for the first time by Max Weber in his 1905 essay 'The Protestant Ethic and the Spirit of Capitalism' (1989).

This original victory for liberal capitalism in the late nineteenth century, made possible by the 'growing strength of industrialism' as Polanyi put it (1957:7), set in place the economic and political conditions for the ramping up of the processes of social acceleration. The growing technological complexity and geographic spread of liberal capitalism at this time contributed to the earliest significant and connected phase of time-space compression, the point in history at which the first Empire of Speed was launched.

The concepts of 'time-space compression' and 'time-space convergence' were formulated by social geographers such as David Harvey (1989) and Nigel Thrift (1996) as closely related concepts through which to analyze the experiential, temporal and spatial effects of powerful and all-encompassing industrialization. As Jeremy Stein (2001:106) explains, the former term refers to 'the cumulative effects of historical improvements in the speed of movements of goods, service and information', whereas the latter concept describes 'the sense of shock and disorientation such experiences produce'. The Age of Capital saw this process take on global and epic proportions. Stein (2001:108) goes on to note that:

> Without doubt, the cumulative effects of technological change, especially over the past century and a half in the field of transport and communication, have been impressive. This was particularly the case in Europe and North America during the nineteenth century when the introduction of railway, telegraph and steamship services radically reoriented geographic and temporal relationships. 'The annihilation of space and time' was a

> common mid-nineteenth century phrase used to describe the experience and significance of these changes. Karl Marx, writing in the 1850s, used similar terminology to describe the improved transport and communications for the circulation and reproduction of capital. In the *Grundrisse* he wrote that 'while capital must on the one side strive to tear down every spatial barrier to intercourse...it strives on the other side to annihilate this space with time, i.e. to reduce to a minimum the time spent in motion from one place to another' (Marx, 1973:539).

The tearing down of 'every spatial barrier to intercourse' also inevitably led to an acute increase in the frictions and tensions between national-capitalist states. *Competition* is what Marx is here referring to as both the motor of capitalism as an inherently expansionary system, and the generator of the contagion of soft speed technologies that shrink space and time. During the last quarter of the 19th century, competition between capitalist states took on new and dangerous military dimensions, because states had the means of making war as an alternative to economic competition and large corporations did not. As the Age of Capital morphed into the 'Age of Empire' in Hobsbawm's epochal historicity, then so too did the dromocratic propensity for conflict between states became more acute. The 'speed of events' such as the 'scramble for colonies' in the late-19th and early 20th century became such that the oft-quoted line 'the inexorable march to war' became a truism as the time for reflection and working through of issues and *projecting their possible consequences into the future* became truncated in direct relation to the quickening pace of the economy, of technological development and a politics that was increasingly suborned to the exigencies of national capitalism. Such an *internationalizing* system thus laid the basis for humanity's first—and by that time inevitable—global war.

Social Democracy: a brief renaissance of reflection over speed

Guenther Roth (2003:264) has written that: 'Many contemporaries failed to anticipate the catastrophes that would befall [classical] liberal capitalism in the twentieth century, and once these occurred, they did not believe in its recovery'. In 1944 when the *Great Transformation* was published, Polanyi (1957:250) also shared the view that no recovery was likely due to a 'congenital weakness' liberal democracy suffered i.e., '...*not because it was* [an] *industrial* [society] *but that it was a market society*' (emphasis in original). This 'congenital weakness' resides in:

> ...the conflict between *the market and the elementary requirements of an organized social life* [that] provided the century with its dynamics and produced the typical strains and stresses which ultimately destroyed that society. External wars merely hastened its destruction. (p. 249) (italics added)

Again using the temporal historical framework it is possible argue that in no small part the 'typical strains and stresses' of society that led to the world wars of the twentieth century were between the incompatible speed rates between a capitalist market economy and 'the elementary requirements of an organized social life...'. The former thrives on speed, whereas the latter functions on planning, an historical perspective, a future perspective, debate, reflection and through conscious effort to resist valorizing time to the point of fetish. The 'congenital weakness' becomes in part a temporal one and is in the inability of the liberal polity to keep in check the speed propensities of capitalist industrialism and the failure to democratize and to decelerate the core functioning of a system that self-organizes around the competition-speed-profit cycle.

The large-scale destruction wrought by two world wars and the immeasurable human misery that is contained under the technical and economistic sounding rubric the 'Great Depression', had shocked and traumatized a couple of generations in the 20th century. An effect of this was that the myths of the automatic beneficence of the liberal democratic/capitalism conjunction could no longer so easily be sustained, as Hopkins and Wallerstein noted. Indeed by 1945, just after Polanyi published his book, a great transformation in society was already underway. Millions of people all over the world had had enough of the effects of 'free market democracy'. In the immediate post war era there was widespread pressure for change, especially in Europe, where much of the structural, economic and social damage had been done. A then-novel experiment called 'social democracy' was very much on the agenda in the shape of governments such as the British Labour Party, the Scandinavian social democratic parties, and the socialist, Christian Democratic and Social Democratic movements of continental Western Europe. These were powerful forces for change that could not be ignored. The will of the people, expressed through their representatives coming to power, had decided to take more of an interventionist role in the volatile operations of the market, with the intention of instilling more predictability and planning into the 'elementary requirements of an organized social life'.

It is interesting to consider that this experiment in social democracy was in many ways a revisiting of the idealistic core of Enlightenment thought. Elements of the 'irretrievably fading' Enlightenment heritage that Max Weber (1989:182) had noted at the beginning of the century were now back in vogue through political thinkers and politicians such as Konrad Adenauer and Ludwig Erhard, as well as influential economists such as John Maynard Keynes. Indeed, Keynes had written his classic essay 'The End of Laissez-Faire' in 1926 during the interwar years, which presented the case for a return to the ideals of classical liberalism and criticized the distortion of Enlightenment ideals by the instrumental demands of commerce. He wrote that:

> ...individualism and *laissez-faire* could not, in spite of their deep roots in the political and moral philosophies of the late eighteenth and early nineteenth centuries, have secured their lasting hold over the conduct of public affairs, if it had not been for their conformity with the needs and wishes of the business world of the day. They gave full scope to our erstwhile heroes, the great businessmen.

Social democracy from 1945 until the middle of the 1970s thus acted as a brake pedal on the speed of *laissez-faire* liberal democracy. And in retrospect the period can be seen as the apex of what democracy has been able to achieve within the historical context of its embedded temporalities (As good as it gets?). The previously unfettered competition-speed-profit cycle that had estranged capitalism from its liberal democratic origins were placed to a significant degree under the regulation of the state in a whole host of important countries. Of course it is not claimed here that big capital was *controlled* by the state, insofar as the state unilaterally decided the direction and substance of the economy, but simply that in the post war era, capital, labour and government agreed, after the trauma of the three previous decades, that a new form of 'partnership' would be in all parties' interests.

In such an environment the 'elementary requirements of an organized social life' could once again be *reflected* on and *acted* on at something approaching the appropriate speed when government took a lead role in the process. There would be the time to do it, because it was not the time it would take that mattered, so much as the 'ideals' themselves (Flora, 1986; Patterson, 1997). And so beginning in North America, Western Europe and Australasia, social democracies began to transform their respective constituencies—this time with the majority of people being positively affected. Mass welfare states were projected, planned and created, a new social context where the 'elementary requirements' in terms of health, education and social security were developed and improved under state auspices. The speed of a more socially oriented democracy was able to function in a more-or-less loose synchrony with the speed of a more socially responsive capitalism.

This is not to say the post war decades were a phase of slow down and inertia—far from it. Economies boomed, factories hummed and heavy industries roared with activity and change and creativity. Indeed as late as 1963, at a Labour Party conference, British Prime Minister Harold Wilson could boast about the 'white heat of the technological revolution' that was sweeping Britain and much of the Western world. This now-famous phrase was a politician's rhetorical flourish, of course, but it did signify something important, something we can only appreciate if we analyze it through the converged prism of speed, technology, politics and capitalism. Wilson's words reflected a confident polity, where social-democratic principles could lead; where political institutions could grant to themselves the power to *control* the pace and nature of technological development—speed where necessary and caution and prudence where it was deemed appropriate. Institutional politics could achieve this because it was in partnership with corporate capital and organized labour. The speed of society, we can now see, was under the control and auspices of these major stakeholders who could all see self-interest in the processes of social democracy as long as political prerogatives, profits and jobs were not threatened.

Still, the 'technological revolution' that Wilson could point to, made for a process of rapid transformation. Industry and society accelerated because of the conjunction of spatial and temporal factors that proved amenable to capital, labour and government. First of all there was the space (geographic and social) for capital to expand into. Much of Europe had been devastated, physically, by six years of war. The rebuilding of lives, communities, cities and their infrastructures gave enormous scope and spur to economic activity. Through wholly political initiatives such as the 1947 US Marshall Plan, where over 12 billion dollars (an astronomical sum at that time) was channeled to 16 countries, new and expanded markets and millions of jobs were created. The process snowballed massively as economic activity was stimulated and more new markets and yet more jobs were generated in the emergence of what would become the post war consumer society, a society where consumption and the production of culture would merge into an increasingly ubiquitous process of Fordist commodification (Slater, 1997).

The time factor was propitious, too. During this period of economic boom, the so-called 'golden age' where profits and prosperity were rising across the board, and where unemployment and inflation were low, the pressures on the core logic of speed and competition were eased. Life could and did speed up during the 1950s and 1960s but this was through the consumer emoluments of Fordism as a total way of life. Wilson's 'white heat' of technological innovation, as it turned out, merely envisioned a consumer revolution, a rather pallid form of Enlightenment 'progress' in the form of the purchase of a washing machine for the first time, or a new car or television, or *entrée* into the growing 'jet set' who traveled abroad. The expanding consumer-based economy meant that the pressure to ruthlessly speed up productive processes, to cut costs and to plough vast amounts of investment into technologies that would allow this to happen was not acute. Moreover, state subsidies and tariffs protected the inefficient and slow from the full force of competition, and powerful unions would fight to protect uneconomical jobs and outmoded work practices. As long as profits flowed to those who mattered most in the most powerful elements of the capitalist system, then the management of the economy in partnership with dilatory unions and stolid government departments could be tolerated. The spatial growth of the post war globalizing system, in keeping with the moderating temporal influences of social democracy was, according to Thrift and Taylor, 'dramatic but controlled' (1982:1).

However, geographic space is limited and the inevitable pressure on profits began to be felt, as a result of overaccumulation, as early as the mid-1960s. Fordism as a historically novel mode of capitalist organization began to reach its spatial and temporal 'frontiers'. As competition became more acute, both big government and powerful unions were identified by big business and by a growing number of neoliberal intellectuals (who were influenced by thinkers such as von Hayek) as the obstacles to continued prosperity. In particular it was perceived that their general resistance to the introduction of technological innovation, especially computer-based technology in the production process was a salient 'rigidity' in the Fordist system (Kolko, 1988).

The speed of capital (increasingly being referred to as the 'mobility' and flexibility of capital) began to make itself an issue once again (Pierson, 2001). Although this was seen at the time as an issue of straight profitability by big business, it was an issue that quickly morphed into one of identifying all the perceived obstacles to the proper functioning of the capitalist economy as a whole—from overpaid and underproductive workers, to 'ideological' unions and governments obsessed with red tape—all of which were supposedly strangling the natural entrepreneurial spirit of the hard-pressed capitalist. Throughout the 1960s this pressure cooker economic situation moved inexorably towards crisis point. The final reckoning came with the oil shocks of the early 1970s. The global economic crisis that ensued precipitated the 'neoliberal turn' and signaled the rise of what we now call globalization (with the new focus on speed, efficiency, cost-cutting, computerization and interconnectedness that rapidly evolved into the 'network society') and the effective end to the tentative post war social democratic enterprise (Pierson, 2001). From now on the 'pace of events' would have its locus in the volatile neoliberal marketplace instead of within the more measured and considered (and slower) ethos of social democratic thinking and planning.

Power and inertia in liberal democracy

I have tried to argue that the politics of liberal democracy has its own timescape, a temporality that is inherited in large part from the contexts of its creation as a political mode of thought and agency. It is important to note that this form of organization and its slow-beating temporal rhythm are prone towards an inertia that contrasts sharply with the restless dynamism of liberal capitalism. It also needs to be understood that this is a structural problem with liberal democracy, with outcomes, as we shall see, that are not necessarily positive for citizens of the liberal democratic system. But what are the essences of this inertia?

A major obstacle in the way of a vital and dynamic polity that is able to respond flexibly to the speed of liberal capitalism is, somewhat ironically, *power.* On a macro level, politics is of course vitally concerned with power and its exercise: the power of legitimacy, the power to marshal the forces of society that can be directed toward the realization of specific ends, the power to enact enforceable laws that help shape culture and society, the power to punish those who transgress those laws, and the power to see off threats from antithetical power—be it from hostile neighbors or from domestic insurgency.

However, Ben Agger argues, apropos the failure of the liberal democratic state to sufficiently create the conditions that positively transform the lives of its people, that: 'Why things do not change for the better is summarized simply enough in words like "power" and "domination"...'

(1992:47). What Agger implies is that power tends to *concentrate* and its 'weight' acts as the means for domination by the powerful and the privileged, the 'power elites' in C. Wright Mills' classic study (1970). This has by now become a fairly mundane observation, except that when we consider Agger's quote more closely its meaning can present itself as being profoundly temporal. It is the *speed* of classical liberal democracy that provides the tendency for power to concentrate. Its *relative slowness* emerges from the contexts of its essential dynamics, such as the fixity of the written word, the physicality of face-to-face meetings, the embedded rhythms of talking, debating, and the making of speeches, the working though of arguments and the coming, eventually, to the issue of recommendations, provisional findings and final conclusions. Gilles Deleuze declared that 'it is in the nature of power to totalize' (1997a). This will and tendency to totality is an adjunct to relative slowness. In the operation of liberal capitalism, however, speed tends towards shortcuts, missing steps and not reading the signposts properly. Power has no time to consolidate or to stabilize; those who wield it must be continually on the move, to seek new opportunities, to stay ahead of the pack lest they be caught and devoured by it. By contrast, slowness and deliberation in the traditional political process tends towards taking in the whole picture and seeing the totality as its proper range of perception and intended effect.

The inherent speed of classical liberal democracy generates other important tendencies that have kept its processes relatively slow. One is the tendency towards the 'iron cage' of bureaucracy, famously theorized by Max Weber as an intrinsic function of modernity (1989:181). Political bureaucracy is of a unique kind. It has a hierarchical 'weight' that has its own beat and tempo. For example in the departments of state services, anywhere in the world, a visit to a place such as an unemployment benefit office, or a passport office, is to feel the 'weight' of the slow-moving pace of power and bureaucracy. It is to experience individual helplessness in the face of what is often an unresponsiveness and unfathomableness (a process, as Neil Postman (1993:116) argues, that is actually *enhanced* by computer technology). Indeed, Franz Kafka, through the surreal travails of his character Josef K, created a literary genre around the speed of bureaucracy. Such deadening administrative systems backed by state power, allows the inefficiencies, costly illogicalities and expensive delays that could not be tolerated for long outside the political field.

The other tendency that follows from slowness and the concentration of power in liberal democracy is the inevitability of oligarchy. In 1915 Robert Michels gave sociological expression to this in his classic theory of the 'iron law of oligarchy' that he developed in his book *Political Parties* (1966). In it he found that all political parties and organizations eventually develop oligarchic power structures, and their individual leaders (over time) are liable to exhibit autocratic tendencies—'*le partie c'est moi*' syndrome, as Michels termed it. In other words those in power tend to like to keep it, and will devote time and energy to develop the bureaucracies and mechanics of power that will insure that this occurs. The social Darwinism of the wider political system, indeed, *compels* parties to do this if they are to function effectively within its logic. As a consequence they become less and less responsive to pressure from 'below' from the rank-and-file of the party and from society generally, where the effects of constant change are felt most acutely. The time factor has not been directly applied to Michel's theory by those who have used it as an analytical framework, but Dieter Rucht perhaps makes the point unconsciously when he writes that politics 'tend to become more centralized-bureaucratic and more moderate in their actions *over time*' (1999:27) (italics added). Paradoxically, perhaps, as William Scheuermann observes, the bureaucratisation and centralization of power is the form best suited to *reacting* to the narrow imperatives of acceleration. He writes:

> Ours is an increasingly high-speed society, and high-speed society places a premium on high-speed political institutions: the widely endorsed conception of the unitary executive as an 'energetic' entity best capable of acting with despatch means that social acceleration often promotes executive-centred government and the proliferation of executive discretion, while weakening broad-based representative legislatures as well as traditional models of constitutionalism and the rule of law.

And yet as power concentrates into the hands of political oligarchies and bureaucracies, the inability to change and the temptation to stick to old ways, to become more 'moderate' or 'conservative' are unavoidable. The time of politics doesn't stand still, but it becomes relatively slow, regulated as it is by the fixity of the institutions of politics itself, its processes, protocols and conventions. 'Over time', a mechanical routinism acts so as to dull the political imagination and confidence to the point where real alternatives to political organization are no longer seriously considered.

Oligarchies and bureaucracies, of course, develop in other fields, such as in business. The tendency is strong here too, but the market itself, through the dynamics of competition (where is exists adequately) continually exposes these inefficiencies. If ongoing (and speed driven) restructuring and regeneration do not occur 'over time', then inefficiencies will result in falling profits, lower competitiveness and, ultimately, bankruptcy. The global economic crises of the 1970s was therefore, as we can see from the temporal perspective, a reckoning with speed, space and time that was inevitable in the social democratic entanglement with capitalism.

Today, as Zygmunt Bauman (1998) argues, power within corporations in the environment of neoliberal globalization tend toward 'dephysicalization' and 'weightlessness'—that is to say, toward provisional power formations that synchronize more readily with the functional speed of capital. The ephemerality of power in the network society—the fact that power is always 'elsewhere', as Scott Lash argues (2002:75), that it is volatile and mercurial in the private sector—is something that is regularly validated in the marketplace. This is conspicuous in the constant waves of 'creative destruction' that ensure that those corporations and their bureaucracies and oligarchies can suddenly cease to exist if they are not fulfilling expectations of short-term profitability. Witness the rise and fall of Enron Corporation, WorldCom, Tyco, Arthur Andersen and the ironically named Long Term Capital Management, to name only a few. These examples show how unstable is the basis of power within neoliberal capitalism. By contrast, the autocratic political leader can be assassinated, be voted out, or be deposed by cabinet or by the people. However the slow-moving power of the *political system* is geared to simply incorporate the next political party or proto-autocratic leader into its intrinsic rhythms and tendencies.

The economic crisis of the 1970s was also a crisis for politics. The cumulative effects of the post-war experiment with social/liberal democracy served to artificially constrain the natural tendency towards ever increasing speed. The 'resolution' to this crisis was therefore also fundamentally a political one. It was the voluntary drawing back from participation in the market process by the institutions of the state. The speed of the market was allowed to function again at levels that were generated from the freedom to use such technologies as deemed necessary to promote efficiency and rapidity in production, distribution and consumption. And as we have seen, information and communication technologies were most suited to this process and its momentum kick-started the information technology revolution that continues apace today. The brief flirtation with social democracy was transformed—through the rise to power of market oriented ideologues such as Margaret Thatcher and Ronald Reagan—into what came to be known as a *neoliberal* democracy, with the ideologically-charged prefix 'neo', indicating the new-found confidence in market processes, the 'natural' way to cure the 'rigidities' and 'sluggishness' of the post-war experiment in social democracy.

Politics *is* able to move faster in the neoliberal economy. But as Scheuerman observes, 'high-speed political institutions' evolved primarily in response to narrow economic pressure. Executive government was intended to act with haste in periods of emergency or crises, such as in war or economic collapse. It is a branch of government, additionally, that would have obvious advantages in its notional ability to act decisively in other times too, and to override or ignore the slower tendencies of the judiciary or legislature. It would follow that executive-centered government has attractions for a neoliberalized economy where speed is of the essence. It gives the appearance of 'administrative efficiency' in a world where efficiency is all; efficiency thus overrides democratic procedures because they are increasingly proving to be too slow in a fast-paced world. A common response for perceived *instances* of executive high-handedness is for the critic of such actions to decry it as imperious or arrogant. A consequence of this is that the concentration of power into the hands of the executive is often represented as (and reduced to) the work of individual personalities who crave power for its own sake. The pressure for the polity to act quickly as a *general neoliberal tendency*—and what this may signify in a systemic sense *vis-à-vis* the relationship with capitalism—is thus missing from such predictable analyses. This is a general neoliberal tendency we can illuminate with an example.

In July 2006 the American Bar Association (ABA) released a report on the huge proliferation in the use of an executive instrument called a 'signing statement'. These are statements that the president attaches to a Bill sent up from congress that he has signed into law. The Bill passes, and on the face of it, it seems like there has been no executive veto or interference. However, the 'signing statement' written by the president says, effectively, 'I pass this bill into law, but intend to ignore it and do not feel bound by it'. The ABA noted in its report that '... it was Ronald Reagan who first used signing statements 'as a strategic weapon in a campaign to influence the way legislation was interpreted'. And since

1980 there have been more 'signing statements' attached to passed legislation than at any other time since the inception of the American republic. Incredibly, and with almost with no publicity, George W. Bush himself has signed 800 of them since coming to office in 2000, whilst only 600 in total had previously been signed since the time of George Washington (ABA, 2006). Rule by executive-centered government may be even more pronounced in developing countries where democracy has been only unevenly established. It is in countries like these, in Peru for instance, where the executive has always had a strong role, that neoliberalism finds its optimal relationship with government and offers, perhaps, a disconcerting glimpse into a generalized future scenario (Klein, 2007). The *New York Times* reported in 2000 that 'Peru is considered a democracy because it elects a president and parliament. [But] in the five years after an election [in 1996]...the executive has been known to make 134,000 rules and decrees with no accountability to the congress or public' (Scheuerman, 2001).

The timeline of the burgeoning of 'signing statements' in the US is instructive. It follows almost precisely the timeline of the rise of neoliberalism, with the election of Ronald Reagan in 1980. The pressures on George W. Bush were evidently even greater that his predecessors, and executive authority was wielded to a unique degree. Indeed the extent of executive authority of the Bush administrations from 2000–2008 was on full view to the world with the US led attack on Iraq in 2003, a Washington elite-led military adventure that was at its core an expedition to protect the interests of US capital in the Middle East and ensure hegemony over the strategic region of oil production to which the US economy is dependent (see for example, Cockburn and St. Clair, 2004).

Short-termism becomes the norm for planning and legislation, with there being insufficient or no time to properly research, debate and analyse the possible consequences of legislation before it is enacted. Montesquieu's 'checks and balances' are relentlessly weakened in the deregulatory process. Liberal democratic polities in their transformation into neoliberal organizations have left the market to operate and to globalize according to its own unpredictable logics. Market-friendly politicians, overridden or compliant legislatures and government bureaucracies, now make a virtue out of their disengagement from direct participation in the economy. In the early 21st century this has continued to the point where state-owned enterprises and large, organized and influential labour unions are seen as an anachronism, a relic from another, much slower, and consequently less efficient and profitable world. And in a reversal of the ethos of public creation and ownership of the 'the elementary requirements of an organized social life...' that developed through the temporary dalliance with social democracy, 'the provision of [such things as] hospitals, roads and prisons' as George Monbiot notes, '...has [now] been deliberately tailored to meet corporate demands rather than public need' (2000:4–5). Freed from much of the burden of public responsibility and duty, most governments now focus obsessively (and competitively with other governments) on the establishment of the 'right business conditions' that would attract the highly mobile capital flows to their particular countries, and thus enhance their global credentials as a 'business friendly' city, region, or whatever.

Neoliberal democratic polities no longer shape the 'pace of events' as liberal democracies did in the 18th century and as social democracy did during the evanescent post-war phase. Neoliberal politics does not lead but instead follows the vagaries of market competition, with results that inevitably do not favour the masses of ordinary people—the professed *raison d'être*—but global capital. This failure to lead and to plan (and respond to the unexpected) was graphically illustrated in the New Orleans floods of 2005, when the government of the world's most powerful nation was unable to act in the most basically efficient way within its home territory (Davis, 2005).

As Scheuerman puts it, executive-centred government acts on the basis of a permanent 'economic state of emergency' where capitalism and its health are the primary focus (2000:1980). With the abnegation of its historical responsibilities, liberal democracy has sold its soul. It has sold its 'own' times that were the temporal-social contexts of its beginnings and its traditions. It has exchanged them for abstract speed that is oriented towards the nebulous concept of efficiency. Most importantly, perhaps, it has sold its heritage and links to history as well as its stake in the future in exchange for the temporal and ethical vacuum of the constant present.

The need to do things faster has only increased the propensity towards oligarchy and power concentration in political democracies, a process that leads in its turn to further instability and uncertainly within the whirlwind of globalization. The legislative effects are inevitably, as Scheuerman noted, executive-centred, pro-speed, 'anti-liberal' and 'anti-democratic'.

Speed and constant acceleration are now hyper-valorized and geared towards pulsing money, ideas, commodities and information throughout the global system, leaving institutionalized democratic politics and their traditional ethics and principles of democracy and accountability floundering weakly in their wake. In this temporalized perspective, politics, democracy and our already-diminished relationship with time, take on yet more gloomy features in the context of neoliberal globalization and the information technology revolution. To quote Chesneaux (2000:417) once more:

> As the [clock] time continuum is hastily and happily deconstructed, democracy dwindles to the shadow of itself, desiccated, shortsighted both upstream and downstream, reduced to purely functional and functionalist objectives and references. And speed enters the picture; it is expected to provide society with a new, introverted field of activism, as a substitute for the now obsolete horizons of the future as well as for the rich, successive layers deposited by the past.

Our present age is one that Virilio characterizes as a 'dictatorship of speed' that 'increasingly clash(es) with representative democracy' (1995). A major problem is that the clashes are at the abstract level of the *idea* of representative democracy. Our living 'representatives' are almost all under the enthrallment of neoliberalism. Markets and competition and informationalization are, more than ever, prescribed for us as the solution to economic and social problems. Within fewer and fewer realms are the economy and society within the purview of institutional politics. Accordingly, in many of the Anglo-Saxon economies, large swathes of social security, health and education are already commodified and placed on a 'competitive' footing (Monbiot, 2000; Hassan, 2003; Head, 2004). And continual pressure is applied to those more obstinate areas (such as guaranteed unemployment benefits, or the health care safety net) that have not yet been brought under the supposedly purgative effect of the market.

In those countries where the social democratic experiment held out for longer, such as in France, Germany and Scandinavia, neoliberalism (or what thinking people in these countries call the much-feared 'Anglo-Saxon model'), is making inroads in terms of declining union membership, increased flexible working and growing income inequality (Glyn, 2005). These are already high-speed societies; it is just that their governments are backing away with a little less haste from their responsibilities in those residual areas where the market demands a presence. The trend is salient across the world. For example, China and India, where most of humanity live, liberalises and marketizes further each day; the Asia-Pacific region more generally speeds up as governments withdraw (Rao and Mendoza, 2005); and Russian capitalism combines oligarchy, mafia criminality and authoritarianism with complex connections to a high-speed global economy in a unplanned endeavour that no-one controls and no-one can predict.

Virilio's characterization of speed as a 'dictatorship' is perhaps the wrong word. It connotes domination by an individual, a cabal or a régime—something physical that may be focussed on by those seeking to resist or overthrow it. Speed, or the propensity to accelerate is not embodied in anything human—speed *works on* humans in all the negative ways we have seen so far. Speed is an abstract process, an idea of efficiency and competitiveness that has become *dominant* through the centrality of a neoliberal ideology that depends on it. In some ways this makes the problem more difficult as people, individuals, groups and communities have less to focus on. How do you protest against 'the market' when we are constantly told that the market is beneficent, or at least neutral? How do you oppose speed when it is embodied in digital networks that are supposedly built to make us more productive, more efficient and wealthier?

In other ways, however, this domination makes the problem more interesting. It is difficult to think of individuals internalizing the existence of a dictatorship of speed. At some level of understanding dictatorships are recognized as such by those who toil under them, generating at the very least a private hatred and a stubbornness to do only what is necessary. Domination on the other hand is a more benign and subtle process. Accordingly, people can incorporate the domination of speed through ICTs into their lives. They can do it and feel confused about how and why they feel oppressed.

Nevertheless it is possible—through the fact that neoliberalism encourages 'selfempowerment' and 'innovation' to use speed-generating ICTs more effectively—to develop ways and employ means that do not strictly accord with the largely instrumental intentions of their creators. ICTs and the global networks, applications and processes that they are developed to create, in other words, can be used subversively.

In the final chapter of this book the upside in what has been until now a fairly dismal account of the transition from Fordism to post-Fordism is discussed. Here we find the appearance on a global scale of a new politics, or new forms of old politics (as yet uncoordinated and unfocussed) that have emerged as a dialectical response to the processes of neoliberalism and the destruction of the liberal democratic polity though excessive speed.

CHAPTER SEVEN

TIME FOR POLITICS: A TEMPORALIZED DEMOCRACY

The Street finds its own uses for things—uses the manufacturers never imagined.
William Gibson.

It's a story about community and collaboration on a scale never seen before. It's about the cosmic compendium of knowledge Wikipedia and the million-channel people's network YouTube and the online metropolis MySpace. It's about the many wresting power from the few and helping one another for nothing and how that will not only change the world, but also change the way the world changes.
(Time Magazine *on the decision to nominate Web 2.0 'You' as Person of the Year in 2006*).

The possibility of democracy is emerging today for the very first time.
(*Hardt and Negri* Multitude, *2004*)

Précis

This book has argued that we live in a postmodernity, one that has essentially been the effect of the transition from the 'high modernity' of Fordism, to the postmodernity of 'flexible accumulation'. There have been many cultural, social and ontological dimensions to this essentially economistic transition. This final chapter looks at the 'political pessimism' effect of postmodern thought on the Left, an effect Fredric Jameson saw as the 'eliminating [of] any memory trace' that could envision other political possibilities to a dominant neoliberal globalization. The political pessimism that emerged over the 1980s and 1990s as a result of the rise to dominance of the postmodern worldview, led to an atrophying of the Left and a diminishment of the intellectual power of Marxism, of social democracy and of the concomitant analytical framework of political economy with which to reveal some of the most important dynamics of neoliberal globalization—those of computerization and the social acceleration that derives from it.

The lack of such an understanding of the nature of neoliberalism has led also to an insufficient grasp of the contradictions of the neoliberalized global economy. The chapter shows that for all the ideological power of the neoliberal worldview, for all its success in 'eliminating any memory trace' of what was, and an indication of what could be, the severe and globalized depredations of neoliberalism has nonetheless provided the context for a set of formative and nebulous (and globalized) political alternatives. This is a 'technopolitics' that is unified primarily by its anti-neoliberalism. In the network society, technopolitics is the consequence of the 'digital dialectic' that is the operation of capitalism's age-old contradictions carried forward into the 21st century. However, the operation of this particular dialectic does not presage the end of capitalism. The technopolitics of anti-neoliberal global civil society movement does not in fact or in theory represent the arrival of capitalism's 'gravediggers'. The traditional Marxist typologies of class and revolutionary movements evaporated with modernity. Networked society is consumer society to its very core. Anti-neoliberal activists, for example, march in Nike trainers and organize through cell phones. What this indicates is that citizens of the global civil (networked) society want the fruits of new production technologies—and not the overthrow of capitalism *per se*. Nevertheless, techno-political activism is mass and is global and is infused with modern tropes of justice, equality and democracy. As a totality they form a rhizome which exerts an upward political pressure upon the neoliberal elite. This pressure, this democratizing antithesis to the worst excesses of neoliberalism—together with the building contradictions and crises of the neoliberal global economy—are producing the contexts for positive change. These are social, political, economic and technological contexts brought together and made explicit through focusing upon the effects of temporality and speed. This focus throws into relief the dynamics of clock time and network time and reveals these for what they are: human artifacts. One is a mechanical device that began to dominate our conception of time in the 18th century, and the other a computer-driven process that drives our networked lives today.

The book ends by discussing how a reasserted control over our temporal relationships with the clock, the network and with the embedded times of our body and the environment that surrounds us, can herald the beginning of a more inclusive and more democratic networked society and globalized economy. The 'bad' speed of neoliberalism can then be replaced with the 'good' speed democratic control.

Politics and the effects of forgetting

The 20th century was dominated by a particular kind of politics. It was a century where, despite the traumas and the catastrophes, the purpose of institutional politics and the boundary-lines of these politics were comparatively clear. Politics was about contesting for power, the power to shape economy and society. In the earlier part of that century it was still a fairly elite project, where in the industrialized West big business and big politics tended to be inhabited by the same people from the same social class. The tensions inherent in this historical arrangement culminated in the inter-elite World War of 1914–18. The 1920s and 30s saw the blowback to that kind of politics in the form of the extremism of Left and right in the shape of fascism and what evolved into Stalinism. Nevertheless, political power, and who gets to hold it, was the common theme throughout the instability of the century's first forty-five years. Post-World War Two saw a reaction to the chaos of the market and the chaos of war, and the increasingly educated and politically aware populations of the more mature (and war-ravaged) economies demanded a radical reforming of political institutions and political practices. The limits of possibility were subsequently expanded, but the direction was still more-or-less clear, and the project of democracy, in the form of social democracy, was seen to have taken a progressive turn for the better.

The 1960s and 70s was in many ways the culmination of this new found democratizing energy and the political space it created. Within this space feminism, for example, could grow, as could radical democratic politics in the form of the New Left, or the Campaign for Nuclear Disarmament. Moreover, the baby boomer generation that was coming of age experienced a cultural awakening that transformed music, literature, fashion and art, and commenced a new engagement with the environment and the limits to sustainability. This space was socially created, but it was financed by a booming world economy. However, as we have seen, the 1970s saw an economic crisis begin to grow, and this had its inevitable effects upon culture and society. After 1968 a pessimism began to set in for the Left, and he narratives that had recently seemed so self-evident in terms of the path to 'progress', the nature of 'truth' and universality of a particular kind of 'democracy' quite quickly did not seem so unambiguous after all. As economic boom engendered a cultural and political renaissance, then so too did economic crisis provoke a cultural and political decline—a decline that produced a new 'philosophy', an essentially 'uncritical' philosophy that, as coincidence would have it, was rather well-suited to the rising forces of neoliberalism (Norris, 1992). An effect of that decline was that political pessimism brought with it a loss of political memory. This was a forgetting about the historic achievements of democracy and consequent loss of faith in what was possible through the agency of inclusive and democratizing structures. With the passing of time, the generational division meant that as the 1970s moved through into the 1980s and beyond, the new generations, X and Y were born into a high-speed individualistic society where politics were increasingly marginal—serving to dissipate social political memory even further. Let us look at the intellectual context of this forgetting.

Postmodernity and political despair

Postmodernism is discussed here at some length because it is important to my argument, and is important notwithstanding the fact the post-modern debates have largely run out of steam—and have themselves largely been forgotten. It is important because the nature of the post-modern debates and their consequences continue to have an effect and contribute still to the political loss of direction that afflicts our social and cultural world. Moreover, it is a political disorientation that is a result of social acceleration that in turn serves to sustain social acceleration through a lack of critical thought and a lack of transformative political agency.

I have been describing the contemporary period as being 'postmodern', or as being characterized by a spirit or feeling of 'postmodernity'. In the opening part of this chapter I want to show that postmodernity, a worldview that extends back early into the last century, set the intellectual and philosophical conditions for the relative political apathy of today and has exacerbated the increasingly atrophying structures of institutional politics. The definition of postmodernity that has been used should be clear enough, but to recapitulate: postmodernity, for me, has its roots deep in the convergence of neoliberal ideology and the revolution in information and communication technologies. It is a post-modernity that is economic, technological, political and, perhaps most importantly, *ideological*. Through the transformative effect of so-called market forces and the now-vital need for speed, the political, social and cultural underpinnings of modernity are being bypassed in the rush towards an unplanned and unknown future. The ontological certainties, the provisional universal truths, the 'grand narratives' and the sweeping historical temporality of modernity—that is to say, a distinct feeling of a sense of past, present and future—are now blurring and distorting through the force of dynamic economic and technological change that have trespassed into every realm and into every relationship.

Other thinkers did not see postmodernity as stemming from this particular locus or having this particular effect. What might now be called 'mainstream' postmodern thought perceived the times through a different, and often more positive, theoretical lens. Influential theorists of the early 1980s, such as Kenneth Frampton and Craig Owens, framed the analysis in the cultural and aesthetic realms and considered postmodernity as an ending of the modern 'myths of progress and mastery' over nature (Foster, 1983:vii). The focus in the cultural and aesthetic realms in these debates, quickly fed into a process that led to what Antonio Gramsci in another context termed a 'pessimism of the intellect' concerning what modernist politics could achieve. In broad terms, the cultural and aesthetic locus of postmodernism began to seep into other discourses, in literature, in debates on the nature of civil society, feminism, consumer society and communications (see Foster, 1983 for a seminal discussion).

What this expanding perspective led to was a fundamental questioning of the validity of everything that had grown and evolved within the 'myths' and 'masternarratives' of the period of modernity. Above all, the questioning and the new postmodern discourses that began dominate, constituted a retreat from traditional politics and a growing pessimism about their being able to exert a democratic influence over the new and radical trajectories of economy, society and technology that were developing in the late 20th century (Agger, 1992). This abnegation created the political and ideological vacuum on the Left that allowed neoliberalism, and its project of globalization and the revolution in ICTs, to become so dominant.

These postmodern debates, which reached something of a zenith in the 1970s and 1980s, have their genesis in much earlier writing in which we can see clear traces of the time-space thinking that has so much real-world consequences today. For example, the intrinsic pessimism and time-space framing of these very European debates has a clear echo in the works of German theorist Walter Benjamin. In his essay *The Storyteller*, written in 1936, Benjamin comments on the loss of the narrative coherence of storytelling (and thus a fix on 'reality') through the general mechanization and relative acceleration of society. He wrote that:

> The art of storytelling is coming to an end. Less and less frequently do we encounter people with the ability to tell a tale properly. More and more often there is embarrassment all around when the wish to hear a story is expressed. It is as if something that seemed inalienable to us, the securest among our possessions, were taken from us: the ability to exchange experiences. One reason for this phenomenon is obvious: experience has fallen in value. And it looks as if it is continuing to fall into bottomlessness. Every glance at a newspaper demonstrates that it has reached a new low, that our picture, not only of the external world, but of the moral world as well, overnight, has undergone changes which were never thought possible. (...) For never has experience been contradicted more thoroughly than strategic experience by tactical warfare, economic experience by inflation, bodily experience by mechanical warfare, moral experience by those in power. A generation that had gone to school on horse-drawn streetcar now stood under the open sky in a countryside in which nothing remained unchanged but the clouds, and beneath these clouds, in a field of force of destructive torrents and explosions, was the tiny, fragile human body.

Here Benjamin describes the effects of contemporaneous time-space compression on the 'fragile human body'. His focus is on the revolutions in the mechanization of warfare that the First World War had made so horribly efficient and apparent. The shrinking of time and space meant, in the modern war context, was that no one was safe from being enveloped within a growing war zone that had not only spatially expanded its realm of operations, but had also temporally accelerated the speed at which conflict was conducted. Benjamin's thesis also implied certain social and *political* effects. These were picked up and developed in Virilio's 1986 *Speed and Politics* where he wrote that 'history progresses at the speed of its weapons systems'. This was for Virilio, a retrograde development, one which—as early as 1914—spelled the end for the Marxist and Enlightenment-derived project of human emancipation through modernity: 'From 1914 onward, the proletariat's motor—and thus political—power was no match for the [bourgeois] European battlefields' (1986:97).

Virilio's gloom may be seen as emblematic of the growing pessimism of once-progressive French thought that reached something of a nadir in the wake of the crushing of the Left in May 1968. The defeat was, according to Rod Kenward 'a final twist of the modernist, ideological spiral which had conditioned all French politics since the 1920s' (2005:244). The 'postmodern turn' began to emerge in earnest after this period, as disaffected Left-wing intellectuals began to drift away from political activism and toward the realms of thought which dominated postmodern theory—such as philosophy, literary theory, literary criticism and aesthetics (Norris, 1992). The disputes, such as they were, all had a distinctly metaphysical ring about them, and they took place within the pages of weighty books and sober journals, which considered questions such as 'what is/was modernity?' and 'what is post-modernity'? Intellectually speaking this was high-end stuff conducted by the foremost contemporary thinkers. But it was thinking that shied away from the political and economic questions that dominated the time, that is to say, the rise of neoliberalism and the dwindling of any socially progressive alternatives that would challenge it.

Jean-Françoise Lyotard, for example, in his celebrated and much-debated *The Postmodern Condition* that was discussed briefly in Chapter Four, believed that the eclipse of modernity was to be welcomed. Postmodernity (however it came about) represented for him an end to the 'totalitarianism' of the metanarratives of science and technology (amongst others) that gave modernity its rigid and narrow perspective. However, far from seeing the 'postmodern turn' as a release into a new historical phase of freedom, Lyotard averred, somewhat contradictorily, that it was contributing to the widespread commodification of knowledge, and to an equally narrow perception of what it is to be human in the world. This bleak vista was made possible through the speed régime of ubiquitous computerization that was geared toward creating and disseminating the valorized and instrumentalized knowledge required for the capitalist marketplace (1979:3–4).

The allegedly corrupting influence of the post-industrial reliance on digital information systems was further—and even more negatively—analysed by fellow French thinker Jean Baudrillard. For Baudrillard it is a computerized and networked *global media* that plays a leading role in the political transformation to a networked age. Through the hyper-proliferation of the electronic image, in advertising, in movies, in video-games and in the daily signs and symbols of commodity capitalism that suffuse our life, the media create what he terms a 'hyperreality'. Baudrillard's thesis follows logically from the 1968 work by Guy Debord, *The Society of the Spectacle*, where the 'fetishism of the commodity' obscures the dynamics of political power. Baudrillard argues that:

> The fetishism of the commodity—the domination of society by 'intangible as well as tangible things'—attains its ultimate fulfillment in the spectacle, where the real world is replaced by a selection of images which are projected above it, yet which at the same time succeed in making themselves regarded as the epitome of reality (1988:137).

For Baudrillard, the incessant 'projecting' of an image above reality produces a hyperreality, a 'simulacrum' or copy of the real that is devoid of depth and meaning. The fetishism of the commodity through the signs and symbols of the image (the spectacle) have been globalized and supercharged and accelerated by the neoliberal revolution in ICTs. In Baudrillard's reading of the postmodern, the loss of reality and the loss of meaning, leave people atomized and exposed to the stupefying effect of a constant plethora of digital stimulation. The unavoidable implication in Baudrillard's work is that such epistemological trauma renders futile any kind of coherent and progressive politics.

The generally baleful influence of so-called 'French Theory' was powerful and spread quickly onto fertile ground during the 1970s and 1980s. It reached into the Arts and Humanities departments of the Anglo-Saxon academies, bringing 'postmodernism' to the fore as a theory to describe a new era of relativism, depthlessness and meaninglessness. The works of Lyotard, Baudrillard and in a slightly different way, those of Jacques Derrida, acted for many as a conscious or unconscious explanation of the seeming end of the 'project of modernity' and the actual collapse of a socialist alternative in the face of the surging neoliberal tide. In his *Late Marxism*, American theorist Fredric Jameson spoke for a significant section of Anglo-American Marxism that was in its *dénouement*:

> ...late capitalism has all but succeeded eliminating the final loopholes of nature and the Unconscious, of subversion and the aesthetic, of individual and collective praxis alike, and, with a final fillip, in eliminating any memory trace of what thereby no longer existed in the henceforth postmodern landscape (1992:124).

Of course these debates took place in academic journals or in books that sold only to a narrow stratum of academicians. The majority of people in either continental Europe or in the Anglo-American countries were mostly oblivious to the theorizing going on in sundry departments in the institutions of higher learning.

This sealed world was only occasionally broken open to public scrutiny, and then usually as a source of ridicule. In 1991, for example,

Baudrillard published in the British *Guardian* (from the original which appeared in the French daily *Libération*) an article entitled 'The Gulf War Did Not Take Place'. In it he argued that what we 'witnessed' in the war to oust Iraq from Kuwait was not a war, but a *simulation* constructed through electronic mass media. Baudrillard's point was that the global mass media could, in effect, create almost any 'reality' it wished through its control over the proliferation of digital signs and symbols. However, right-wing media, and mainstream media in general chose to ignore this subtlety and ridiculed Baudrillard (and by implication all 'French Theory') for arguing that the war did not occur, that it was all a media confection. This willful or obtuse misrepresentation was held up as 'proof' that intellectuals (French or otherwise) were hopelessly out of touch with the concerns and thoughts of men and women in everyday life. The impression was reinforced in 1996 when Alan Sokal, a leading quantum physicist, published a highly divisive piece of writing in the cultural studies journal *Social Text*. His article 'Transgressing the Boundaries: Towards a Transformative Hermeneutics of Quantum Gravity' appeared to argue—using an excessively opaque version of the language of social theory and cultural studies—the need for a postmodern perspective on the mathematical and physical sciences, a perspective where relativism and subjectivism should be given a legitimate role in the physical sciences. Sokal later unveiled his article as a hoax on the 'anti-science' rhetoric of postmodern thinking. Ridicule was heaped on the editors of *Social Text* who failed to pick up the sham article at the review stage. Sokal's point that one could say almost *anything* within the postmodern discourse and have it taken seriously. It was an observation that was eagerly taken up in the mainstream media as yet another example of how improvidently taxpayer's money was allegedly being spent in the universities.

The political consequences of postmodern pessimism

Except when being encouraged by the media to join in the fun and pour scorn on 'intellectuals' (the term usually had unmistakably cynical inverted commas attached to it in the Anglo-Saxon world) ordinary men and women were sidelined from these debates. And this is a central issue; it fitted neatly with an ideological agenda. The academic obsession with the 'postmodernity debate' missed the point. For a decade and more, it served mainly to divert many critical minds from properly analyzing the locus and scope of the *actual changes* that were rapidly occurring in society. For a mainstream society at the mercy of increasingly narrow media viewpoints, the alleged babble of the Left-leaning academic or postmodern thinker seemingly 'proved' that, for a start, socialism and social democracy was finally and decisively bankrupted. The general internalization of this concept meant that much else which dovetailed with the neoliberal project was able flow: from the idea that 'there is no alternative' to neoliberal restructuring and the natural supremacy of market forces, to the realization that 'there is no such thing as society' only an agglomeration of competing individuals in a market context. Both these utterances came from Margaret Thatcher in the 1980s. Thatcher blazed a trail for neoliberal ideology that quickly dominated British politics and pioneered the global transformation of economy, culture and society. Her close ideological ally, US president Ronald Reagan, spoke for the business and rising political class in that bastion of neoliberalism in terms of market power, when he said in his first Inaugural Address in 1981 that: 'government is not the solution to our problem; government is the problem'.

It is no coincidence that during the early 1990s when the subject of 'globalization' was being seen as the 'next big thing', the postmodernity debates were beginning to peter out. In fact these terms, globalization and postmodernity, were describing the same processes—only the former better describes the cause and the latter its effect. The central 'problem' during the 1980s and 1990s was not one of postmodern ontology and metaphysics; these could be debated *ad infinitum*. The pressing issues were (and still are) economic and political issues, the dominance of neoliberal globalization and the reckless speeding-up of economy and society. As the analytical framework of this book would suggest, this transformation can be seen as due to a technologically and politically driven time-space compression which resulted in the shift from the regime of Fordism (high modernity) to one of flexible accumulation (postmodernity). In this historical materialist analysis, this shift represented the basis on which economy, culture and society became postmodern or, more accurately, neoliberalized and globalized.

The political pessimism of the post-1968 era allowed an academic and essentially arid postmodern debate to obscure this fundamental process. Theorists and intellectuals got caught up in essentially irrelevant disputation and neglected to first analyse the real world transformations that lead to flexible accumulation, before looking at their ontological and metaphysical ramifications. A significant political problem stemming from this was that a Left-wing critique had lost much of its legitimacy in the eyes of a public bombarded by a neoliberal media and (by the 1990s) a politically conservative academy (Hassan, 2003). The analytical power the Left had derived from a historical materialist and flexible political economy approach had become increasingly marginalized during the 1970–90s, not only in the postmodern debates, but also in the succeeding debates on globalization that continued through to the 2000s (Frankel, 2001).

Alternatively, the foregrounding of temporality within an historical-materialist and political economy approach would have revealed that of supreme importance was the *disconnection* of Enlightenment and clock-time based politics from an open-ended continuum of market-driven speed through ICTs. This would have revealed a crisis within the institutions of political liberalism, a crisis where social and economic acceleration effectively made the 'project' of liberal democracy impossible to realize. However by turning away from politics, a French Theory-derived postmodernism allowed the crises of politics to go unchecked and misunderstood.

Into this apolitical vacuum rushed the ideology of the market. Its ascendancy helped to ensure that many accepted as inevitable the disruptive transformation to flexible accumulation that would allegedly bring wealth, efficiency, productivity and profit as its rewards. In this real world context, the question of the erasure of the 'memory trace' that Jameson notes, is therefore not a philosophical one (in the first instance) but an ideological and technological one—with profoundly *temporal* effects. The past is left to be forgotten or to be written from a perspective that naturalizes the constructed trajectories of the market, of science and technology, and of the individualism and relativism of the postmodern (and politically apathetic) present. The future, for its part, is painted as a vague but always-ominous horizon of anxiety, one where surprises will be many, and one where it is the individual who must take responsibility to insure against whatever it may bring.

Unlike the post war baby boomers, generations X and Y have no experience of an institutionalized political culture that has 'naturally' occurring ideological differences—or even generally different policies. Today the ideology of neoliberalism puts up a broad front in the parliaments and in the congresses and senates of the developed and developing world. It is an ideological homogeneity that is reinforced and disseminated by a powerful global media, which like the university, functions primarily as a business like any other (McChesney & Herman,

1997). It is arguable that the majority of younger people sees no political differences and perceives no crises in the archaic and disconnected party political system. And as the mainstream media hardly ever tackle issues of crises within the party political system, then people simply don't consider it. Society appears (contradictorily) to be about the individual, about the need to lower taxes, about mortgage rates and the price of gasoline, school fees and so on. In this context social acceleration continues apace and networked globalization and the supremacy of the market are viewed as ordinary aspects of life.

However, equally arguable is the idea that for others, as society becomes increasingly atomized and the concerns of people are yet more instrumentalized in their orientation, then the apprehensions, the stresses, and the time-squeezes can act as catalysts for reflection, for questioning, and for action from an overtly political perspective. And it is within this liminal space—that is to say, between the promise and the reality of neoliberal globalization—that political hope lies.

The limits to neoliberal domination

If neoliberalism functions too fast to allow it to deliver on the promises it made to people in the 1970s and continues to make today; and if liberal democracy functions too slowly to deliver on its 'historic responsibility' to its citizenry, then what is the alternative? The truth is that no one knows. Accompanied by its empty and largely cynical adherence to the tenets of 'liberalism' and 'democracy', persistent economic and social crisis continue to follow in the fast-flowing wake of the neoliberal project. All the while its ideologues and unthinking followers assure those who still listen that things will be all right if we become yet more efficient, yet more productive and yet more competitive. Millions fall into the daily routine of unthinking conformity, or convince themselves that there probably are no alternatives.

This is of course a form of ideological domination. But it is a highly complex one in the context of a seemingly free and open network society where diversity abounds. Ostensibly, the Internet carries every manner of idea, every topic of discussion and every form of analysis. Websites proclaiming the virtues and necessity of anarchism, or socialism, or fascism, or eugenics, or communitarianism, or almost any topic you care to type into Google, can give the impression of a vibrant, almost subversive digital space, one that undermines the very basis of the neoliberal speed economy. But it isn't. The Internet and its predominantly business networks are monopolized by instrumental uses. These submerge everything else. We are online in the office, the home and in the school and the university to do the 'new economy' work and learning that increasingly consume our waking lives. We spend long hours processing information in our factory-offices, we catch up on the backlog of tasks at home at nights and on the weekends. Children, in preparation for a life in cyberspace, spend more and more time reading, writing and researching on the Internet at school and at home.

The leisurely 'surfing the Internet for fun' trope that is beloved of ICT advertisers as a hook to get people online, actually represents a fairly small part of what people do there. For example, the Pew Internet and American Life Project found that of the 220 million Americans (some 75% of the population) who used the Internet daily (in 2008) only 28 per cent spent some time in idle website hopping. A somewhat larger fraction of this number (39%) accessed the Internet to find mainstream political information and news (Pew, 2008).

It may be overwhelmingly apolitical and business-oriented, but it is not all doom and gloom on the Internet. When not using the business or educational networks, it seems that what people overwhelmingly do at home (and in the office if they dare) is share music and video files through peer-to-peer (P2P) protocols. *Wired* magazine cites research by IT firm Cachelogic, which found that 'P2P applications consume between 60 percent and 80 percent of capacity on consumer ISP networks' (Glasner, 2005). Spending much of your time downloading the latest movie blockbuster or hit songs may be technically illegal, but it is certainly not political or subversive. Indeed, the 'illegal' period of being able to freely swap and download copyrighted music and films is rapidly coming to an end as the business model of P2P asserts itself. This was highlighted in 2006 when the iTunes website run by Apple sold its one billionth 'legally' downloaded song (titled, fittingly enough: 'The Speed of Sound') to an unsuspecting user in the USA. The lucky customer received *ten* iPods and an iMac computer as a prize (Apple.com, 2006). The number of iTunes songs sold tripled just a year later (Apple.com, 2007).

The ideology of the market and the commodity culture clearly dominate the network society and the increasing time we spend within it. It is the Baudrillardian nightmare of domination through signs and symbols writ large. It constructs a super-mediated existence where the 'formal' network of the Internet and its peripherals, with their speed and their commodification and their instrumentalism, seeps into realms well beyond the job or the classroom. According to Thomas de Zengotita, this Baudrillardian hyperreality through speed creates a kind of 'numbness', a lack of feeling for the real and the actual, for the essence or the originary level in daily processes. This is analogous to Kuipers' metaphor of 'constantly treading water at the surface of change' (2000:12). In his 2005 book *Mediated* de Zengotita composes a playful piece of writing to catch the rhythms of this numbness:

> All about time. Crunch time. Time to get a life. Busy, busy, busy being numb. You don't need a Blackberry, you need a chief of staff. Quality time. Down time. Even the food is fast. Real time. She runs marathons too. The end of the day (p. 175).

Zengotita gets rather more serious a few pages later when he gets to describing the effects of these stultifying dynamics. He goes on to observe that 'our state of numbness' is created by being:

> ...swamped by routine activities. The old-fashioned superficiality of routine blends seamlessly with the new superficiality, the surface quality of ubiquitous representation—and this hybrid accelerates constantly, as you take on more and more. (...) The result is a simulation of reality convincing enough to pass for the original, for most of us, most of the time. It is only when the ultimate reality descends on us in the form of a tragic accident, illness, death or a miraculous recovery, the birth of a child—only then does that simulation stand revealed for what it is (p. 186).

According to Baudrillard this speed-driven dynamic results in 'the purest and most illegible form of domination' (1988:130). This domination, however, is not effected through a simple dichotomy between 'reality' and its 'simulated' and accelerated representation. Super-mediation means that there is no 'ultimate reality' as Zengotita puts it, except for those events such as the death of a friend, or a personal illness that can cut through the thickets of signs and symbols. Domination stems instead from the fact that within the process of super-mediation there exists little personal autonomy, few areas of actual diversity that is not premised on the capitalist commodity logic.

Within this logic, 'diversity' and 'choice' are potentially limitless—as we see in the Internet itself. But these are constrained, ultimately, by the simple necessity of what sells and what can be made to sell.

Domination. It is a word that feels like the weight of a load on your back, or an oppressive mental force that obliges us to think this way or that, or to be this or be that. Domination can seem to operate so successfully—through a well-oiled ideology—that it functions to the point where it becomes invisible (Eagleton, 1991:47). We no longer feel its effect on us, we no longer discern its point of production or the heaviness of its presence—we come under its spell, in other words. The ideology of religion functions in such a fashion, and the ideology of neoliberalism functions in ways that approach this point of invisibility. In politics and in the mainstream media, ideas such as market forces, free trade and liberal democracy are routinely referred to as if they are as natural and life affirming as the air we breathe.

Whatever one thinks of their modalities of analysis, the writings of Lyotard and Baudrillard undoubtedly make the forms of domination in postmodern society readily apparent. But they also lend themselves to a politically pessimistic reading, to what Christopher Norris describes as a 'postmodern-neopragmatist drift' (1992:85) where domination becomes an inescapable (and increasingly invisible) part of the postmodern condition.

However, the term domination can also connote a complex and subtle process. Indeed, a more nuanced understanding of it can in fact reveal the process of domination as actually *a sign of vulnerability*. Such a perspective can identify gaps in what appeared to be the solid edifice of the system. Raymond Williams gives us such a way to think about domination and its operation. He argues that for a power to contemplate 'domination' it must presupposes a resistance, some active force to be kept in check or held in place by a greater force. Williams goes on to observe that:

> ...however dominant a social system may be, the very meaning of its domination involves a limitation or selection of the activities it covers, so that by definition it cannot exhaust all social experience, which therefore always potentially contains space for alternative acts and alternative intentions which are not yet articulated as a social institution or even project (1979:252).

Using Williams' injunction, these spaces of alterity and reflection can be utilized to show that religion, for example, has no answers, and functions mainly to deaden the capacity for rational/critical thinking. Similarly, neoliberalism, for all its dominance in the media and in the business and government spheres, cannot possibly function to the point of invisibility because unlike religion it *forces* itself daily into our lives. It does so through the inevitable and inexorable contradictions that arise between the idea and the reality, the incongruity that emerges between the promise and the product itself. For many millions across the world the inconsistency makes itself felt in ways that can be uncomfortable or irritating—like another tax increase in an environment where government and regulation are supposed to be minimized. More often the contradictions can be very sharp, and the effects a good deal harsher. Jobs can disappear, economies can be wrecked, and families and lives can be torn apart in innumerable ways through the stresses and anxieties of living in a free-market economy where 'restructuring' is a permanent and palpable and painful feature of life.

Neoliberalism's particular vulnerability is that it cannot be sustained as either an economic system or a political ideology. The contradictions of open-ended acceleration and a spiraling consumption without end (neoliberalism's principal time-space contradictions) guarantee that it will fail—but they don't guarantee that what will follow it will be any more sustainable or rational. This is where the immediate challenge lies for a progressive politics. Before looking at the signs of change and the options for political change it is necessary to illustrate some of the primary contradictions of the current economic dis-order.

Capitalism and its contradictions

In their book *What is Philosophy?* the rather more politically committed French theorists Gilles Deleuze and Felix Guattari argue that: 'all concepts are connected to problems without which they would have no meaning and which themselves can only be isolated or understood as their solution emerges' (1994:146). This is almost a straight paraphrase from an even more politically activist Karl Marx who wrote in the *Preface to the Critique of Political Economy* that:

> ...mankind always sets itself only such tasks as it can solve; since, looking at the matter more closely, it will always be found that the task itself arises only when the material conditions for its solution already exist or are at least in the process of formation (1975:182).

What these quotations argue, in simple terms, is that ideas evolve in a continuum, not in isolation. Ideas are connected to their immediate social environment, which in turn are indissolubly linked to earlier ideas and their social milieu. It is a classically dialectical mode of thought which argues more broadly that no society or mode of production or *dominant ideology* can break down and its historical effects be transformed until they have first begun to develop the *contradictory* forms and processes that are inherent in its internal relations. More crudely put, as was often his wont, Marx reiterated this point in *The Communist Manifesto* by writing that 'What the bourgeoisie... produces, above all, is its own gravediggers' (1975:46). At the present time this dialectical and historical materialist analysis is rendered practically invisible by the 'postmodern neopragmatist drift'. But when it is rendered and utilized, many economic and political developments that are routinely seen as marginal, or obscure, or unconnected, take on greater political, economic and social significance.

Neoliberalism—capital's latest and most comprehensive manifestation—is replete with contradictory forms (see for example Galtung, 2005; Jessop, 2007). These stem from the same intrinsic problems that have always been associated with its operation. Today, however, these have been intensified and extensified by wave after wave of ICT-driven globalization that envelop the entire planet. In human terms the impact of free markets and the pursuit of growth and profit at almost any cost, was articulated as I touched on in the introduction in the *2005 Report on the World Situation* published by the UN Department of Social and Economic Affairs (UNDP). Its findings were devastating, but thanks to an institutional global media that lines up squarely behind corporate interests, it barely raised a flicker of interest when it was released. The *Report* noted first of all that the much-trumpeted political and economic commitments regarding the overcoming of global inequality that were outlined at the 1995 World Summit for Social Development, were 'fading'. It goes on to observe that:

> In spite of the compelling case for redressing inequality, economic and non-economic inequalities have actually increased in many parts of the world, and many forms of inequality have become more profound and complex in recent decades (UNDP, 2005).

So much for the decades old neoliberal guarantee that the 'level playing field' and the democratizing magic of market forces would bring prosperity to all. The ongoing economic pressure for a reduction in tariffs in the pursuit of 'open markets' has had a devastating effect on poorer and less developed countries that simply cannot compete with powerful multinationals (Bello, 2005:129–179). And within the richer economies themselves, workers in skilled or semi-skilled manufacturing or IT jobs that can be done more cheaply in China, or India, Russia, or Mauritius, or Guam, or wherever, find themselves, in their millions, being reduced to working feverishly in the booming service industries on lower pay and in more tenuous circumstances (Roberts, 2005). Under neoliberalism it is clear that inequality runs deep and wide and corrodes social cohesion in New York or London as much as it does Abidjan or Mumbai.

A more worrying contradiction for CEOs, for Wall Street analysts, and for the occupants of the boardrooms of the world more generally, is that of overproduction. This is linked the economic effects of inequality we just mentioned, and to technologically driven speed. The need to continually increase the rate of the production of commodities is an ineffaceable aspect of capitalism. It reduces the cost of goods and services and allows the producer to undercut rivals. ICTs have been key to the speed-up in production in all industries under neoliberalism. This can have superficially positive outcomes. For example, Western consumers are able to buy the mountains of cheap goods that run off the factory assembly lines in, say, China or Malaysia. The result is that most of us have more commodities crammed into our lives than anyone would have dreamed possible even a couple of decades ago. Three or four television families are nothing special now. As are two cars in the suburban driveway. And laptops, desktops, cell phones, MP3 players, DVD players, CD and DVDs discs, set-top boxes, PDAs, Blackberry's and so forth, clutter our houses, offices and take up increasing amounts of landfill space.

The expression 'the consumer society' is not even a cliché any longer. We have internalized its logic through saturation advertising. We consume but never feel sated, only dissatisfied and frustrated (Barber, 2007). Notwithstanding the ultimately false promise of advertising that we all recognize at some level, most of us simply take for granted (even in the large stratum of poor and unemployed in the West) capitalism's first rule of never-ending domestic consumption. This habit is prefaced on another exquisite contradiction: we are told that consumers must be 'confident', i.e. we must buy more and more to keep the economy strong. But we are also cautioned that we must save more—for our mortgages, for our retirement, for periods of unemployment, for our children's education and so on. However, it is impossible to do both at an increasing rate. So we tend to carry on consuming because it is easier and more pleasurable in the short term. Living on a day-to-day basis means that increasingly we find ourselves with fewer savings or no savings at all. For example, fully twenty-seven per cent of people in Britain do not have a penny in the bank (Toynbee, 2005). So the 'free market' answer to this is for debt-financed consumption to take up the slack, causing levels of personal indebtedness to soar to crisis proportions (Padgham, 2007).

Spiraling consumption of course has its limits. People need and can afford to buy only so many TVs and cars—yet production has to go on, and indeed it *must continue to expand and to accelerate* if individual capitalist concerns are to survive. The car industry is a classic example of what the industry calls 'overcapacity'—or too many unsold products piling up. In 1998 academic and activist David McNally cited industry figures which estimated that 'global excess capacity in autos... is around 21–22 million cars... roughly a 36 percent overcapacity relative to world markets, the equivalent of 80 efficient state-of-the-art plants' producing at full capacity (1998:27). In 2004 an industry consultant told Reuter's news agency that carmakers in the US alone were sitting on some four million surplus vehicles worth US$100 billion (Reuters, 2004).

There is an overproduction in almost everything. If there isn't then entrepreneurs eagerly will jump in to exploit the opportunity until the market is once more glutted and crises ensue. The logic is as old as the capital relation itself. The ICT industries are allegedly the quintessence of efficiency and 'new economy' thinking, but the logic here is exactly the same. Economist Robert Brenner writes that:

> In 2000 no fewer than six US companies were building new, mutually competitive, nationwide fiber-optic networks. Hundreds more were laying down local lines and several were also competing on sub-oceanic links. All told, 39 million miles of fiber-optic line now criss-cross the US, enough to circle the globe 1566 times. The unavoidable by-product has been a mountainous glut: the utilization rate of telecom networks hovers today at a disastrously low 2.5–3 per cent, that of undersea cable at just 13 per cent (2003:54).

As Brenner goes on to argue, this example and countless others like it show that the market, left to its own logic, certainly does not know best. In fact the permanent tendency towards overaccumulation demonstrates that the neoliberal 'solution' to the self-same crises in the 1970s was no solution at all. Indeed, it points to the inescapable conclusion that a 'managed' economy is *more* efficient because through regulation, tariffs, and consultation with the stakeholders involved, the crises of overaccumulation can be deferred for much longer period of time. High Fordism can thus be seen as an unconscious experiment in the social control of time and speed. The post war Fordist 'managed' boom was the longest uninterrupted period of prosperity in history. The neoliberal 'rule', by contrast, has been a highly compressed three decades of turbulence, restructuring, social dislocation, atomization, and the widening of levels of inequality across the world.

A digital rhizome emerges

In the face of such reality people have either hunkered down, hoping for better times, or they have grasped the initiative and begun to fight back against the depredations and contradictions of neoliberalism. The growth of alternative political forms to institutionalized neoliberalism since the early 1990s takes us to another contradictory logic. It is one that has so far been under-appreciated, but it is one that is nevertheless working to produce its own antithesis. This logic has its roots in the development and uses of ICTs, comprises, to borrow phrase from Peter Lunenberg (2000), a 'digital dialectic' that is emerging to challenge the neoliberal order.

Under neoliberalism the ICT revolution was to be all things to all people. It would free the economy from its choking red tape and would liberate the individual who had been held in the binds of statist serfdom since 1945. Neoliberalism, if it subscribes to nothing else, subscribes to the tenets of individualism over collectivism. It was the core thinking of the Thatcher and Reagan revolutions of the 1970s and 1980s and on into the 1990s. It has been their enduring legacy into our own time. Its ideas flowed into government policy formation and into business strategy. Big business took to these Hayekian principals with alacrity (Frank, 2000:34–35). The ideology of individualism worked positively in the minds of business leaders, and this is most clearly evident in the adoption and uses of new information technologies on three specific levels.

First, is that ICTs and the spread of free-market individualism made possible the post-Fordist economy of flexible accumulation. This effectively divided and atomized the old Fordist tendency toward collective action in the workplace, thereby fatally undermining the organization of labour that grew and flourished in the modern period. It was the inculcation of a social-collective mentality amongst the working classes that under postwar social democratic Fordism in particular was able to function as an effective resistance to the introduction of new 'productivity tools'. Second, is that it is cheaper to leave people with relative autonomy to explore and learn and innovate with ICTs. They are able to do not only what is required of them (and more, always more) in their jobs, but they are also able without prompting to follow their own interests and passions, to experiment at home as well as the workplace, to write code, to tinker and fiddle, to discuss better ways of doing things and to develop new processes and applications for computing. This is the sort of basic interest and passion that allowed the ARPANET researchers of the 1950s and 1960s create the fundamental infrastructure of the Internet. Third, for the capitalist who provides the environment for keen and willing individualists to pursue their genuine passions, they provide also the potential for further economic growth. It was in such an 'open' environment that entrepreneurs such as Bill Gates, Steve Jobs, Jeff Bezos or Google's Larry Page and Sergey Brin could emerge to build the new industries in which more profit-making opportunities could be explored, and new markets exploited.

So are we left with a world of ICT-driven individuals, where collective political action is no longer possible? A world where 'individuality' is at our psychological core, with 'society' being simply an expression of market transactions? At one level this would appear to be the case. Individualism and its economic effects surround us. It has left a trail of destruction and misery and feelings of powerlessness amongst those who had to pay the cost of the anti-collective, anti-social and anti-humanist policies of the last quarter-century (Davis and Monk, 2008). At another level, however, the atomization of societies has not irrevocably scattered people to the ends of the Earth. Look at any period of history and see plentiful evidence that, whatever Margaret Thatcher and others aver, people need to be social. Cooperation, to paraphrase Marx with a dash of Darwin, is part of our 'species essence'. It would follow from this that if you try to create and manipulate a world of individuals by giving them powerful communication tools, the results are likely in many cases to be something other than you expected.

It is a fact that neoliberal globalization and its dependency on competent users of ICTs in its pursuit of 'efficiencies' and profit, has created a global strata of *technologically savvy people*. These are the millions who can see the potential uses of ICTs through their expertise with them, but who also still think critically and are able to express a measure of commitment to the Enlightenment tenets of Reason, justice and democratic political participation at all levels of society through their expert uses of information technologies.

Since at least the early 1990s, prior even to the Internet becoming mainstream, individuals and collectives had been active in developing and articulating new political languages through innovative use of ICTs

(Hassan, 2004:100–138). As John E. Young of the Worldwatch Institute commented 'As of mid-1993, thousands of environmental activists and organizations around the world [were] using commercial and non-profit computer networks to coordinate campaigns, exchange news, and get details on the proposals of governments and international organizations' (1993:21). The effects of networked neoliberalism were global and so, in fairly short order, were its networked and global responses. Famously, the 1994 'Zapatista' rebellion in dirt-poor Chiapas Province in Mexico, showed not only how deep went the logic of neoliberalism, but also how far the recognition of the need for a coordinated (networked) political resistance to it had penetrated. The Zapatista struggle (especially in respect of its inventive use of ICTs as a weapon) was picked up by sections of the media and became a *cause célèbre* in the struggle against the worst aspects of free enterprise. For example, in July 1996, in a bold show of international solidarity, over 3,000 protestors from over 40 countries around the world, gathered in Chiapas to take part in what came to be known as the 'First Intercontinental Gathering for Humanity and Against Neoliberalism' (Flood, 1996). Delegates at the meeting called for the creation of the 'Intercontinental Network of Alternative Communication.' Their spokesperson was Subcommandante Marcos, the leader of the Zapatistas, whose statement (cited in Leal 2000:7) read:

> Let us start a communications network between all our struggles, an intercontinental network of communication against neoliberalism, an intercontinental network for humanity. This intercontinental network will seek to tie together all the channels of our words and all the roads of resistance. This intercontinental network will be the means among which the different areas of resistance will communicate. This intercontinental network will not be an organized structure; it will have no moderator, central control, or any hierarchies. The network will be all of us who speak and listen.

These activists came together just as the 'Internet revolution' really got underway and the movement in general gained much momentum from the growing presence of ICTs in everyday life. The Internet had been given a massive boost around the middle of the decade with the giveaway browser software that came with Windows 95 and with the ultimately doomed Netscape Navigator that could be downloaded free from the Internet itself. IT businesses boomed as the dotcom bubble began to grow unchecked. The process of 'ubiquitous computing' that Mark Weiser and John Seely Brown (1997:5) describe became a reality as almost every industry began to believe their own hype and come to view computing as the ultimate solution for every problem. IT corporations went on large-scale 'recruitment drives' to get people online—and millions accepted and signed up to new companies called Internet Service Providers (ISPs). People began to surf the net and work or consume as it was intended for them to do; but many technologically savvy activists also began to make their own non-instrumental connections. These 'intercontinental networks' were explicitly political and anti-neoliberal and were forming into a loose and amorphous (yet densely connected) movement.

We saw its globalizing beginnings emerge from the International NGO Forum that was the 'alternative summit' to the elites' 1992 Rio Earth Summit. The 1990s was the decade that saw technopolitics grow on a world scale, with the *événements* of Porto Alegre, Seattle, London, Genoa and other cities snowballing into what became known as the World Social Forum (WSF). Emerging formally in 2001 at Porto Alegre, the WSF is a worldwide agglomeration of activists of every kind, whose main reason for coming together is a shared antipathy towards neoliberal globalization. North American Christians, animal liberationists, right-wing French farmers, Italian anarchists, British socialists, Brazilian rubber workers, Australian new age hippy's, Filipino trade unionists, South African miners and many others use ICT networks to swap stories of their plight under neoliberalism and seek or suggest alternatives. This diffusive coalition comprises what is in effect a *global civil society in cyberspace*, an embryonic substitute to the perceived bankruptcy of their own local civil societies in Brazil, in France or wherever it may be.

This global civil society is a grassroots movement in a genuine sense. It is an emergent energy that encompasses, politically, all that neoliberalism is not and cannot be. Its participants look to the collective as opposed to the individual; and they look to democracy instead of the marketplace. What is most important about it is that it is only made possible through information and communication technologies. Through them they become, potentially, a force, a culture and an evolving political language that communicates globality as a human solidarity in place of an economic opportunity. This virtual movement also represents a new relationship with time and space. It is a *virtual political autonomy* that is in elemental form. Its teething problems are many and some might possibly prove fatal. Michael Hardt attended the second Porto Alegre WSF in 2002 and wrote the following for the *New Left Review*:

> The [Porto Alegre] Forum was unknowable, chaotic, dispersive. And that overabundance created an exhilaration in everyone, at being lost in a sea of people from so many parts of the world who are working similarly against the present form of capitalist globalization (…) The encounter should, however, reveal and address not only the common projects and desires, but also the differences of those involved—differences of material conditions and political orientation. The various movements across the global cannot simply connect to each other as they are, but must rather be transformed by the encounter through a kind of mutual adequation (…) What kind of transformations are necessary for the Euro-American globalization movements and the Latin American movements, not to become the same, or even unite, but to link together in an expanding common network? The Forum provided an opportunity to recognize such questions and differences for those willing to see them, but it did not provide the conditions for addressing them. *In fact, the very same dispersive, overflowing quality of the Forum that created the euphoria of commonality also effectively displaced the terrain on which such differences and conflicts could be confronted* (emphasis added).

The final sentence encapsulates the dilemma that faces the global civil society movement. The many and varied effects of neoliberal globalization has brought together an enormous range of people who want to resist it. However, if the effects of neoliberal globalization are so diverse, stretching across so much of the world, how are people able to confront their differences sufficient to not only resist neoliberal globalization effectively, but also to change it? It is far from clear that anything approaching an organized and plausible alternative can emerge from this 'overabundance' of political energy. Nonetheless it is a political force that cannot stay still nor be ignored. Its energy-filled composition means that it will continue to evolve and change as neoliberalism does likewise. Indeed this 'digital rhizome' that had been growing during the 1990s had already reached the point of self-awareness vis-à-vis the importance of ICTs tools for the creation of alternative times and spaces. As Naomi Klein wrote in 1999 in the *New York Times* of the Seattle demonstrations against the policies of the neoliberal-dominated World Trade Organization (WTO):

> This is the first political movement born of the chaotic pathways of the Internet. Within its ranks, there is no top-down hierarchy ready to explain the master plan, no universally recognized leaders giving easy

> sound-bites and nobody knows what is going to happen next. But one thing is certain: the protestors in Seattle are not anti-globalization; they have been bitten by the globalization bug as surely as the trade lawyers inside the official meeting. Rather if this new movement is 'anti' anything, it is anti-corporate, opposing a logic that what's good for business—less regulation, more mobility, more access—will trickle down into good news for everybody else (…) The face-off is not between globalizes and protectionists, but one between two radically different visions of globalization. One has had a monopoly for the past ten years. The other has just had its coming-out party (2002:3–6).

In its diversity, collective resistance to the power of market forces and the perceived whittling away of the rights to free speech and free communication has taken other forms that are not so directly political but nonetheless add to the totality of networked resistance. The 'open source' movement, for example, has catalyzed a global stratum of those who write computer code, or are activists who are concerned about questions of free speech in cyberspace. These are the tech-savvy architects of the formal network society, but they tend to bridle at what they see as its overly proprietary and commercial nature (Robins & Webster, 1999:255). Lawrence Lessig is one such campaigner who recognizes the dangers of a cyberspace that is being privatized by corporations through rapacious property laws. A lawyer and academic by training he advocates a much more free and creative cyberspace, an open and public domain that is not trammeled and constricted by copyrights and the 'user pays' culture. His books have been profoundly influential in creating the pressure for what he terms a more democratic 'regulability' of cyberspace. Indeed his 2004 book *Free Culture* was made available for free download at www.free-culture.org. In the Introduction to his 1999 book *Code and Other Laws of Cyberspace* Lessig argues for control, but control of a specific kind:

> Liberty in cyberspace will not come from the absence of the state. Liberty there, as anywhere, will come from a state of a certain kind. We build a world where freedom can flourish not by removing from society any self-conscious control; we build a world where freedom can flourish by setting it in a place where a particular kind of self-conscious control survives.

In other words freedom cannot be left to a speed-filled cyberspace where there is no one in control, where there is no direction but where the market takes it, and where no account is taken of those who cannot pay. The 'open source' movement which was developed by Richard Stallman in the 1980s, merged neatly with which Lessig's thesis that people, not corporations, own ideas (see Stallman, 2002). Written by tech-savvy libertarians, such as Linus Torvalds who created Linux, 'open source' software was 'free' and programmers were able to make modifications and improvements to it in what is called the 'copyleft' system. This 'general public license' (GPL) is at the core of the open source movement and permits users to sell, copy, and change copylefted programmes. The finished product can then be copyrighted. However, you must pass along the same rights to sell or copy your modifications and for others to possibly change them further. You must also make the source code of your modifications freely available. With potentially hundreds of thousands of programmers and hackers continually modifying and (one would think) improving the code, the GPL system acts as a sort of Darwinian 'natural selection' where applications that best suit their contexts or environments are the ones that are most successful in meeting the actual needs of people. (Moody, 1997).

Cross-fertilization is rife in the activist network ecology. And the lines between street-based technopolitics, the freethinking libertarians who spend much of their time writing code, and constitutional lawyers who agitate for a creative digital commons are blurring to invisibility. For example Lessig wrote in *TechnologyReview.com* in 2005 of his visit to that years' 100,000–strong World Social Forum, again in its Porto Alegre 'base' in Brazil. He described a 'sprawling collection of wooden huts, connected by a canvas spread across their roofs'.

> This was the free-software lab. To the right, there was a training room, with more than 50 PCs arranged along long tables. At the far end was a large screen, where 20 to 30 kids were watching an instructor explain the workings of some video-editing software. Every machine was running free software only—GNU/Linux as the operating system, Mozilla as the browser, and a suite of media production software, most of which I had never seen on any machine anywhere.

This little scenario captures the authenticity of forms of technopolitics. It depicts self-motivated political action that is driven at some level of consciousness by retrofitted Enlightenment ideas of liberty, justice and the ineradicability of the public sphere. All this, remember, is made possible through a neoliberal-derived expertise with digital communication tools. In his article Lessig ponders over the political ethos which makes possible this (in some ways absurd) coming-together of neoliberalism's antithesis. Drawing from Richard Stallman's Free Software Movement, Lessig perceives the Porto Alegre political and technological driving force as being guided by the Movement's four principles 'geekily numbered starting with zero':

(0) The freedom to run the program for any purpose;
(1) The freedom to study how the program works and adapt it to your needs;
(2) The freedom to redistribute copies so you can help your neighbor;
(3) The freedom to improve the program and release your improvement to the public, so that the whole community benefits.

Lessig concludes by observing where this energy, knowledge and commitment are going:

> Consider how the kids in Porto Alegre think about remixing. They remix culture with words, certainly. But they want to build the capacity to remix more than words. They hope to use computers to remix culture. For most of us, computers are a way to type fast. But for most of them, computers will be a way to speak, using sounds and images, synchronized or remixed, to make art or remake politics.

What these trends indicate is that that the democratic spirit has not been completely eradicated. Its cultural ethos remains, even if its political power has been diminished. And so the anti-globalization movements, free software promoters and the growing aesthetic movement that uses digital art as a new form of language, are *not necessarily political*. Indeed the haziness and diversity that Hardt describes takes much of the edge and the focus off any political possibilities. All is not lost, however. There is a different trend in the rise of blogging, a network practice that does have definite (and often effective) political orientations. It is necessary to devote some time to blogging as it has been described by the Institute for Interactive Journalism as a networked form of 'citizen journalism' that 'lets people from all walks of life take the media into their own hands' and has the potential to transform politics as we know it (KCNN, 2008). It is thus potentially a form of control for people and therefore a means toward political empowerment—something very much lacking in our neoliberal high-speed world.

A political economy of blogging

The fact that blogging has some kind of political future was evidenced during the 2008 Democratic and Republican Conventions in the USA. Here, for the first time officially accredited bloggers attended these hugely important meetings, alongside the mainstream media, many of whom had also jumped on the blogging bandwagon (Fairbanks, 2008). What this signified is that the spectrum of potentially counter-systemic ICT social media users now covers a broad and growing span—from the geeks in Porto Alegre, to the independently-minded code writers who produce 'copyleft' software, and on to the political nerds who opinionate to the blogosphere from their bedrooms or from party conventions.

From its emergence in the early 2000s as a result of technical innovations such as Web 2.0 software that allows for 'real-time' interactivity through web browsers, blogging, like social networking, has gone mainstream. Although seeming to plateau near the end of its first decade of existence, blogging is nonetheless phenomenally popular. The Internet tracking company Technorati estimated that by early 2007 there were 72 million individual blog sites, which is more than the entire population of France or Britain (Cheng, 2008). Blogs are so simple to create and maintain that, inevitably, they will reflect every whim and obsession that people have. Many blogs are devoted to scouring the Internet for weird and wonderful things: links to pictures, streaming audio and video, stories, new gadgets and so on. This trend might be classified as 'general interest' blogging where people can read, comment or ignore at their leisure. However, many, if not a majority of them, would go under the rubric of 'culture, news, and politics'. This realm of the blogosphere is even more chaotic and dispersive than that of the emergent global civil society. With thousands of blogs sprouting weekly, political comment, ideas and analysis can range from the first-rate to the vacuous; and from the liberal humanist to the neo-Nazi racist. There is no ethic or set of rules that exist to uphold standards—and if there were, they would be impossible to supervise or enforce. And who would have the right to pronounce on universal principles across a global diversity anyway?

The point is that they are being embraced by millions of people who spend their time creating them, reading them and thinking and acting on what they have created and read. In other words blogs are having a discernible affect on both the political process and mainstream media whose institutional structures they threaten—or at least are perceived to threaten. Repressive régimes such as those of Iran and China get very uneasy at the thought of free and unregulated communication emanating from within their borders. And with good reason: it was estimated in 2005 that there are about 50,000 bloggers in Iran alone (Garton-Ash, 2005). One of them, Arash Sigarchi, was jailed for 14 years for what a BBC report termed 'charges of spying and aiding foreign counter-revolutionaries' (BBC NewsOnline, 2005). Mainland Chinese authorities, for their part, have been busy blocking would-be Chinese political bloggers from reading overseas blogs or creating their own. Moreover, it was reported (in a blog) that the blocking capability was supplied to the Chinese government courtesy of US Corporation Cisco Systems (McKinnon, 2005).

The growing complexity and interconnectedness of political blogging is apparent in the rapidly growing sophistication and level of organization within an ostensibly unordered blogosphere. Most political blogs have sidebars, or what are termed 'blogrolls' where readers can view links to dozens, sometimes hundreds, of other blogs that share the same interests. At a still higher level of organizational complexity, clusters of like-minded bloggers form in cyberspace to create political pressure groups. One such group is the Paris-based Reporters Sans Frontières, or in English, Reporters Without Borders (www.rsf.org). At their website they provide free access to a publication called *Handbook for Bloggers and Cyberdissidents*. This manual gives plenty of advice on how political bloggers can get to work. Information such as 'how to set up and run a blog', 'how to blog anonymously' and 'technical ways to get around censorship' are offered. Julien Pain contributes an essay 'Bloggers: The New Heralds of Free Expression' to the handbook and in it he writes that:

> Blogs get people excited. Or else they disturb and worry them. Some people distrust them. Others see them as the vanguard of a new information revolution. One thing is for sure: they are rocking the foundations of the media in countries as different as the United States, China and Iran (2005:5).

Reporters Without Borders see blogging as the basis for a potential revolution in political journalism. They see it functioning as a kind of journalistic rhizome where honest and truthful reporting has the ability to undermine the corporate media and the political status quo whose interests it primarily serves. As Pain goes on to argue: '...blogging is a powerful tool of freedom of expression that has enthused millions of ordinary people. Passive consumers of information have become energetic participants in a new kind of journalism...' (2005:5).

This new kind of journalism was brought rather dramatically to the fore during the lead-up to the Iraq war in 2003. In the weeks and months prior to the widely expected US invasion in March of that year, the 'Baghdad Blogger' whose blog name was 'Salam Pax', animated the blogosphere and the wider institutional media with his accounts of daily life in the Iraqi capital where journalists were unable to operate freely. After the invasion he kept posting to his blog, dear_raed.blogspot.com. So popular did his blog become, especially his description of the 'shock and awe' phase of the US bombing, that he was hired by the *Guardian* newspaper to write a fortnightly column (McCarthy, 2003). Much blogging is from a liberal, Left-of-centre political perspective. However, there is no shortage of diametrically opposed blogs. Again during the 2003 Iraq war a blog called blogsofwar.com attracted a large following for its alleged revealing of the 'truths' that the Left-leaning 'liberal press' would not print about the war.

There are now numberless political blogs crowding cyberspace. Many simply repeat what is said in other blogs, and so therefore it is easy to get the impression that bloggers are a collection of politics nerds who spend their time reading each other's postings—many of which would refer to the same news items anyway. I shall say something shortly on what has been termed the 'echo chamber effect'. However, people in power actually read them and are influenced by them. Some particularly enthused politicians have their own blogs and increasingly come to see them as an effective way to reach a wider audience. Authoritarian governments, as we just saw, certainly read them and are prompted to curb or destroy what they see as an incipient parallel political universe that could grow into a threat to their prerogatives.

What are we to make of all this political action? Blogs are modes of online production. They create forms of knowledge that can question institutional knowledge and disseminate alternative views. However, given that anyone can write anything on these sites, does this make them useless, or at the very least suspect, in the pursuit of ideas that could change the world? Moreover, speed appears to be valorized here as it is in the neoliberal economy. So is blogging just another instance of the pervasiveness of speed? Looked at coolly, is this ostensibly alternative domain, simply another realm of information production, and a 'data trash' creator that bombards and disorients us still further?

These caveats raise substantial and legitimate concerns. However, to put the best possible inflection on what blogging represents, it can be said, to begin with, that it is a *consciously political* use of ICTs. Moreover, as the basis for the production and distribution of political ideas, political blogs may be the nascent platform for the formation of a global civil society movement that represents an overtly grassroots form of control over new relationships with space and time. Simply put, blogging potentially amounts to the political and creative *reestablishment of social control over time, space and speed.* How so? Politically speaking, what bloggers are potentially able to do is to *hold* (in virtual space and time) ideas that have been obliterated or marginalized through the effects of the speed-economy of globalization. These network spaces can thus become *retentional areas* where thoughts and ideas develop. This is in stark contrast to the currently monolithic 24-hour news cycle where what was said and published yesterday evaporates as the media move on to other issues. Politically committed bloggers, however, can keep an issue going for as long as the blogosphere feels that it is important enough. It is in this time-space framing that we can see the latent power of the blog to retain cultural and political memory, which is a prerequisite for any form of effective and positive resistance to postmodern 'post politics'.

In a diversity of circumstances and at multiple levels of awareness, people are now able to realize that the computer, mobile communications and the networks that bring these erstwhile business tools to life, have uses that can be resistant and subversive and furnish the means for autonomous political agency. Political actors who have been marginalized from the institutional dynamics of democratic processes, or who have found them to be ineffectual, are discovering through their own technical application, spaces and times of alterity that are able to be developed to work against the dominating pressures of the neoliberalized network. The ability for nearly 'everyone to be an author' and have a political voice that can reach millions is a unique development in human communication, one made possible by instrumentally oriented ICTs being turned to other purposes.

Thinking about blogging in this way is to project a best-case scenario for the capacity for the political blog to transform political processes and institute new ones. But how effective have blogs been in concrete terms? In other words what does the evidence indicate? This is a key question in the context of this book, because it concerns a technology that holds out so much promise for so many people.

The evidence seems to be that political blogging has real-world effects at two distinct levels where there exist two functionally separate spheres of institutional political sophistication. At the first level, political blogs are recognized as a powerful tool for subversion. As Jacqui Cheng observes, the areas in the world where blogging has taken off fastest, is in authoritarian or newly-democratic societies, those with just enough freedom and just enough GDP to allow its use to spread—countries such as South Korea, Taiwan and China (Cheng, 2008). Democratic structures in countries such as these are relatively new, and evolve in the context of a globalized economic system. In China in particular, blogging is growing in tandem with tremendously rapid general Internet access. China is of course only formally democratic, and in practice the one-party state jealously guards against any threat to its power. Freedom of political speech is therefore severely curtailed in mainstream as well as in online media (Reporters Without Borders, 2008). Accordingly, in Mainland China most blogging is restricted to non-political topics such as lifestyle, culture, music, movies and so on. However, it seems to me that a way of explaining the fact that so many Chinese choose go on line to blog about virtually anything apart from political topics, is that at the very least it signifies a yearning for a freedom of speech in any way they can get it. In this sense blogging about movies, for example, can be seen as a form of political engagement in that it constitutes the exercise of a democratic freedom. Where this can lead to is an open question, but what the practice does is to inure Chinese citizens into having an opinion that they can legitimately express. At the present time this acts as a vital safety valve for the government in a country that has deep connections with the global economy, but it is a profound contradiction within Chinese capitalism, and is one that is ultimately untenable.

The second level is in the context of the mature democracies of the West, in the Anglo-American context especially. Here overtly political and critical blogging is widespread and growing. It reflects a diversity of political opinion and a level of free speech that Western democracies take for granted. However, notwithstanding some of the work independent bloggers do in exposing corruption, hypocrisy, keeping an issue alive, fact-checking and so on, they represent no threat to neoliberalized democratic institutions. The unavoidable conclusion is that there is presently no insurrection coming from the blogosphere, no political 'vanguard of a new information revolution' as Julien Pain put it (2005). We see this illustrated, for example, in the craving by some of the most popular and influential independent bloggers to be accepted into the mainstream through their willingness to participate and be 'credentialed' at formal political processes such as the 2008 US conventions (Fairbanks, 2008). *Incorporation* into the larger traditional structures of the political-media system is the striking symbolism here. Striking too is the parallel dynamic of *commodification*, such as when prominent blogs like the liberal *Daily Kos* take advertising. By this act they willingly insert themselves into the network society cash nexus and subject themselves to everything that becoming a business entails. As a platform for genuine free speech and fearless opinion, and for the propagation of new ways to conduct politics, the dual processes of incorporation and commodification strictly compromise any such radical political pretensions.

Moreover, the potentially important role that independent political blogging can play, with its ability to exert some control over time and space in the development of new ideas, of political critique, and of being able to form the basis of new and global political structures for a global age, is hampered by the very digital technologies that give them existence in the first place. For example Cass Sunstein, in his analysis of the political efficacy of blogging, describes what he calls the 'echo chamber effect'. In his 2001 book *Republic.com* Sunstein makes the point that sometime soon in the future, technology will have 'greatly increased people's ability to "filter" what they want to read, see, and hear... [and] you need not come across topics and views that you have not sought out. Without any difficulty, you are able to see exactly what you want to see, no more and no less' (3). It is a new kind of media use that threatens to make redundant any general interest media that cover a whole range of topics in the one newspaper or magazine. Sunstein has a name for this media consumption. He calls it '*The Daily Me*—a communications package that is personally designed, with each component fully chosen in advance' (7). He notes that this kind of technology was hailed by many as a triumph of individuality, convenience and control. Sunstein sees also that this could be viewed as a form of consumer power—the 'growing power of consumers to filter what they see' (8). And ostensibly this does seem like a good thing. You read and see what you want, when you want. No nasty surprises as might happen on old broadcast television, where graphic vision of, say, a car bomb going off in some far off city, with many innocents killed and injured is unexpectedly shown. In this new kind of media consumption, serendipity, or chancing across an interesting article, or hearing a point of view that you were not previously familiar with, becomes increasingly less likely, too, as filtering software becomes more sophisticated. As Sunstein put it, such a development is a false kind of social power that puts private control in conflict with public democracy. The shared experiences that help constitute a 'mass society' where people have a similar perspective on the world and fairly similar forms of knowledge through which to inform this common worldview are in danger of degenerating into a kind of social fragmentation within cyberspace (10–13). Through the technological ability to be exposed only to what you want to be exposed to, opinions, views and ideas ring as if in an echo chamber. As Sunstein observes: 'New technologies, emphatically including the Internet, are dramatically increasing people's ability to hear echoes of their own voices and to wall themselves off from others' (49). Moreover, there is the tendency to listen out only for 'louder echoes of their own voices' (16). This presents a major problem as far as a vibrant and diverse democratic functioning is concerned. Fragmented communication, ghettoised communication, 'niched' communication, leads to a narrowing of opinion. We may feel 'free' and secure within our own digital bubble, but as Sunstein (50) argues:

> Freedom consists not simply in preference satisfaction but also in the chance to have preferences and beliefs formed after exposure to a sufficient amount of information, and also to a wide and diverse range of options. There can be no assurance of freedom in a system committed to the 'Daily Me'.

Writing in 2001 at the dawn of the blogosphere, what Sunstein terms 'collaborative filtering' can be seen as a pre-blogging term for blogging. However, in a 2007-revised edition of *Republic.com*, Sunstein takes up the issue of political blogging more explicitly, and devotes a whole chapter to the subject. He writes that political blogging is but a small percentage of the vast and growing total blogosphere, but nonetheless it 'seems to be having a real influence on people's beliefs and judgments' (2007:138). According to Sunstein, there are genuine benefits to be had from blogging in the political realm. For example, thousands of keen-eyed bloggers can act as fact-checkers on the claims of politicians, or the media, or corporations. Political bloggers are also able to highlight issues and force them onto the institutional agenda, issues that might otherwise have been forgotten or buried in an ocean of information. However, as yet, the evidence of the blogosphere's ability to influence the agendas of the public stage is, Sunstein concludes, 'all too little' (146). He believes that the wiki-based technologies of blogging simply make the echo chamber effect much more efficient. Political bloggers, he observes, are 'primarily interested in cherry-picking items of opinion or information that that reinforce their preexisting views' (143). Recall here my description of 'blogrolls' where like-minded bloggers advertise each other's sites.

Sunstein's main point is that as a forum for 'deliberative democracy' the blogosphere does not work. The echo chamber effect, he argues, fragments the public sphere and polarizes political opinion. To strengthen his case he cites the 'Colorado Experiment' from 2005 whereby in a controlled laboratory situation the polarization of political opinion within selected groups became strongly evident. The experiment consisted of ten groups of six American citizens, with each group consisting of people with either 'liberal' or 'conservative' opinions. They were asked to discuss between themselves three of the most contentious issues of the time: same-sex marriage, affirmative action, and whether the US should sign the Kyoto climate change agreement. Participants were asked their opinion after fifteen minutes of deliberation, and then asked to try to reach a public verdict prior to the final anonymous statement. The results indicated that in 'almost every group, members ended up with more extreme positions after they spoke with each other (60–61). So not only was the polarization effect strengthened, but also the opinions within the polarized groups tended to become more radical. Sunstein suggests that 'it is entirely reasonable to think that something of this kind finds itself replicated in the blogosphere everyday' (145). He cites further evidence that suggests that the blogosphere is indeed comprised of discrete political ghettoes: a study conducted during the 2004 US elections found that out of the 1400 blogs surveyed, fully 91 per cent of the links (the 'blogrolls') were to like-minded sites (Sunstein, 2007:149).

Blogging, or general online communication, can put millions of people onto the street as we saw in the positive perspective—or, conversely, it can become a highly efficient echo chamber where opinions and ideas and strategies for political action are discussed in an increasingly narrow context and where the tendency is towards ever more polarized and extreme opinion. As technological expressions of politics, or 'technopolitics' what can we take from this positive-negative summary? Unfortunately, it is not possible to give definitive answers or clear-cut predictions. Why? Because it is in the nature of the media—and in the nature of society—to militate against such a neat packaging of postmodern political reality. This is even more the case in the context of social/technological acceleration. For example, Wikipedia tells us that although blogging was fairly widespread from 2001, it did not become mainstream until 2004, when easy access to wiki software created the explosion in user generated content. In the space of just a few years it has made an impact, intellectually and practically, as we have seen. However, this rapid acceleration should tell us something. The information society is still evolving. It is evolving in ways that we cannot predict because the logic of the information society, the rationale that drives it, is anarchic and based upon the dis-order of competition. Blogging (political and otherwise) rose rapidly to become global phenomena. But if no one had conceived of it a decade ago, who's to say that it won't just as rapidly fall into abeyance over the next few years, as the next new 'killer app' sweeps the planet as the latest techno-fad? Who, indeed, can argue against the *strong likelihood* of this happening? Change is a constant within capitalism, but capitalist change, or what Joseph Schumpeter called 'creative destruction' has become digitalized and is expressed in the information that flows through the network at speeds that increase with every new microprocessor that becomes the latest 'system benchmark'.

Economy and politics are about social power, and power is now attached to the bits and bytes that comprise the network traffic. As Castells puts it, 'relentless adaptation' and the 'multiple strategies (individual, cultural, political) deployed by various actors…' is what constitutes the quest for dominance in the information society (Castells, 1999). Power moves and stops, dissipates and concentrates, but it does so in the context of a constantly moving dynamic, where nothing stands still for very long. Google, for example, is presently the darling of Wall Street and is perched atop of the globalization/ICT heap with shares worth over $700 a piece and rising. It finds favor at the highest political levels, too, a fact illustrated by a British government initiative to have Google involved with solutions for poverty in Africa (Elliot & Boseley, 2007). They have come a long way in a few short years. But Google could easily fall prey to the power flows of an uncertain and volatile economic/political system. 'Confidence' in the Google business model could evaporate in the wake of a political decision or technological development. Who knows? As for the institutional political realm itself, there is precious little sign that it has fallen out of love with a free market, or with globalization based on a free market premise. This means that liberal democracy's self-marginalization from the loci of economic (and hence social/political) power will remain—and probably increase. Global and local political institutions still invest inordinate power in both the market and information technologies as the solutions for societies ills. They do this in the face of very little supporting evidence, and do it at the expense of democracy. Global warming, for example, is possibly the most important challenge that faces us. But it is a crisis that will not to be tackled primarily through the application of science and the mobilization of political will to prioritize action and to take the rational and logical measures necessary. Instead, humanity is to confront impending catastrophe through the application of a market in carbon trading. In other words, we are entrusting our collective future to the irrational and illogical dynamics of capitalist competition (Monbiot, 2007). Another example of government weakness and lack of democratic accountability is the US government's takeover of the mortgage lending corporations Fannie May and Freddie Mac in late 2008 to avert their collapse. By that action US the taxpayer effectively had assumed liability for trillions of dollars of housing loans, monies that wash around a global system, and are subject to global volatilities that the US government has little control over—other than as a local backstop against neoliberal collapse.

The question begs: what keeps this kind of society (global and local) together? *Fear* might be a reason. Fear of the pace of change (and so we tend not to challenge it, even as it accelerates); fear of our economy collapsing (so we don't question 'the experts' and their 'solutions'); and fear that our politicians don't really know what they are doing anymore (Giddens's observation that we live in a 'runaway world'). In an information society governed by the volatility of market competition and driven by the hyperspeed of networked computers, and in the absence of an effective politics at the local or global level the 'fear factor' makes some sense. Such pervasive fear connects with the discussion on anxiety that was identified earlier as a major 'pathology' of speed. Replace the term 'fear' with 'anxiety' when considering the opinion of Tony Judt writing on the logic of networked neoliberalism in the *New York Review of Books*:

> Fear is re-emerging as an active ingredient of political life in Western democracies. Fear of terrorism, of course; but also, and perhaps more insidiously, fear of the uncontrollable speed of change, fear of the loss of employment, fear of losing ground to others in an increasingly unequal distribution of resources, fear of losing control of the circumstances and routines of one's daily life. And, perhaps above all, fear that it is not just we who can no longer shape our lives but that those in authority have lost control as well, to forces beyond their reach (2007:22).

And fear/anxiety is at the centre of the supposed solutions for the crises of Western democracy: neoliberal globalization and information technologies. The fear engendered by the Cold War was at the core of the computing logic that helped create the Internet. And fear of the threat of rising interest rates, of rising inflation, of rising prices for a barrel of oil, of a return to the generalized crises of the 1970s, and so on, is the gel that binds globalization; not promise, or progress or hope. This baleful scene is a far cry from the rather more positive visions of someone such as Daniel Bell who foresaw the possibility of progress carrying the information society forward through knowledge workers and expert systems that were to be 'managed politically' in the context of a functioning and committed democratic polity (1973:18–19). It is a scene that also is in stark contrast to those present-day boosters of the information society, such as Bill Gates of Microsoft, or for Google's Larry Page and Sergey Brin, who assure us that all kinds of social, material and democratic benefits would flow from ubiquitous computing. They see the information society fundamentally in terms of pure technological progress, where new and improved forms of culture and democracy will automatically follow.

I have tried to propose another reality, one that is constructed through the prism of a political economy of speed. In this perspective we see that political blogging is able to carve out a measure of autonomy and agency within the context of the institutional political structures that have been found wanting. However, within the technological paradigm of the neoliberal global economic system, politics is currently weaker than the orbital pull of speed and capitalist imperatives. What this has meant is that for all the hope invested in the power of the blog, it is ultimately destined to be incorporated and commodified—or marginalized—by the overarching logic of abstract market 'laws'. In the pursuit of a new politics to rid us of the pathologies of the speed economy, we need to think afresh. And so to conclude I want to try to develop some ideas on the future of democracy, by taking all that has been said thus far into a consideration of what might be our best options.

Conclusions
Speed and politics: is there a future for democracy?

In the *The Communist Manifesto* Marx and Engels famously declared that:

> What else does the history of ideas prove, than that intellectual production changes its character in proportion as material production is changed? The ruling ideas of each age have ever been the ideas of its ruling class (1975:52).

Expressive of their teleological and materialist conception of history, the authors of the *Manifesto* expected that the contradictions of capital would precipitate insurrectionary revolution, turning the old order on its head. The new and democratic 'ruling class' would then produce its own 'ruling ideas' for a new Communist Age. Their expectations didn't eventuate, but their concept of society's ruling ideas having a specific social-technological provenance still holds true.

Regis Debray, as we saw previously, reminded us of this critical insight in his 2007 essay 'Socialism: A Life-Cycle' in which he argued that the most important material basis for the production and dissemination of the ruling ideas of a society were the 'communication networks' that predominate at any particular time (2007:9). And, since the time of the industrial revolution the bourgeoisie has of course overwhelmingly owned these. Debray's contribution to Marx and Engels was to temporalize this process in the context of what he called the 'graphosphere' where communication networks based upon printing constructed a great periodization, an 'arc of time' that was only superceded with the coming of the digital age. The 'graphosphere' was, as we also saw, a temporal context, a 'timescape' which conditioned the ruling ideas of the beginnings of modernity, the political ideas that gave the project of modernity its dominating temporal rhythms that were arranged around the bass-line of the clock.

Given the perspective of Marx, Engels and Debray on the nature of the production of ideas, and their material context, we can properly ask, what are the ruling ideas of today, in the timescape of the network society? In the economic realm during the industrial revolution, the ruling ideas consisted of the efficacy of liberalism, or markets and of capitalism more generally—and in the realm of politics, the Enlightenment ideas of progress, of universalism, of a brightly unfolding tomorrow and of modernity more generally, held sway. Today, the ideas that govern the economic realm are similar due to the dominance of neoliberalism, however the political ideas are much less salient and shaping of culture and society. Indeed, as we saw previously, the triumph of postmodernism in much of the universities, in the media and in political institutions themselves, has meant that we are much less sure of ourselves politically. The idea of 'progress' is now seen as somehow naive, the future is uniformly dark for most people, a vague a threatening place that we have little control over, and a deep faith in the capacities of democratic institution to make our lives better has been replaced by a cynicism, or an apathy, or a relativism that eschews any 'totalizing' solution for an innately contingent world. Institutional politics, wherever people take it seriously any longer, is a disordered faith wherein the universalism of democracy is a pipe dream, and a practicing democracy is itself a 'vulnerable local achievement' that stands upon permanently shifting sand (Martin, 2007:10).

In politics, the postmodern view of 'anything goes' has in a sense been the triumph of *no ideas*, of no observable and understandable master-narratives that give meaning to our lives and our place within a specific kind of socially constructed world. This is a serious problem for concepts of democracy that evolved and grew in a very different context. It is, however, very good for neoliberalism and the speed economy of globalizing capital. Today, ideas exist in a marketplace of ideas. They become bits and bytes of information, commodities that may or may not sell in a fickle and distracted consumer society—made fickle through the tyranny of choice, and distracted by the tyranny of speed. We have ideas, but in postmodern theory one idea is as good as any other. It is a logic that would argue, for example, that people in the West have no right to say that Chinese must be more like us politically. And so we have a 'Chinese democracy' where individuals can get rich, but are not allowed to question the government; and a Chinese democracy where individuals can now get sick and die due to lack of socialized health care, but no one is allowed to complain about it too much. The Chinese way, like the American way is thus another 'vulnerable local achievement', but it is one that is not necessarily exportable, because in the postmodern way, the rights of the Other are always to be respected. Again this suits globalizing capitalism well, because the differences between local and global can be packaged and sold as choice and difference and freedom.

The whole thrust of my narrative has been that temporality (speed and acceleration) needs to be made salient in order for us to perceive and understand society better. The temporal perspective allows us to see what effects the Second Empire of Speed has had upon the economic processes of capitalism, and in turn upon the political institutions that have historically been so comprehensively enmeshed with it. We see also that through neoliberalism and postmodernity, and through the effects of a networked speed economy, the ideas that sustained liberal democracy and social democracy have become subordinate to the needs of capital. I have argued this point consistently and have discussed at length the social, cultural and political costs of this transformation.

However, there is another cost that I have not given much attention to until now, and that is the *loss of political memory*.

In the constant present of our network society, where accelerating speed and increasing volume of information keeps us increasingly preoccupied by what is immediately to hand, the social and individual memory of political history dwindles in older generations, and is barely able to take form in the minds of younger generations. Collective memory can span generations, but in the postmodern context it is especially susceptible to selectivity—aspects of remembering that are chosen for their political utility by political elites to manipulate and stir local populations (Misztal, 2003:112114). Here, political memory becomes oppressive of democratic processes. This was graphically illustrated in the Balkans in the early 1990s where nationalist demagogue Slobodan Milosevic repeatedly used the symbolic memory of the Battle of Kosovo in the faraway year of 1389 to help inflame separatist passions in the Serbs, projecting a selective memory of their historic 'persecution' that led directly to the brutal break-up of Yugoslavia.

Selective political memory provides no general threat to the current capitalist system in the way that a widespread recovery of the traditions, promises and achievements of liberal and social democracy would. And this point leads us to conclude that it is necessary to get back to some basic principles that were lost in the fizzling out of modernity and the subsequent rise of postmodernity and neoliberalism in the 1970s. The postmodern debates were largely conducted in the realms of cultural theory, literary criticism, architecture and so on. Progressive politics was a primary casualty here because many of its leading theorists never really took the debates as being intrinsically political. The fact that they were, coupled with the diminishing influence of critical social analysis in the universities and beyond, led to a political pessimism on the Left that allowed neoliberalism to flourish (Agger, 1992:51–52).

Across the world, baby boomers are now settling into retirement, and with them go their receding memories of another age of experience of what a form of politics committed to taking the lead could achieve. Many of this generation was born and grew up experiencing a post-war social democracy, where government-implemented Keynesian economic programmes of free education and health care, unemployment and sickness benefits, public housing, rising real wages, progressive taxation and so on, were instigated. These were the basis for Polanyi's 'elementary requirements of an organized social life (1957:249) and constituted a consensus driven social democracy, where the individual was able to be a part of the promise that democracy had held out since the dawn of the Enlightenment.

Generations X and Y have no such experience and therefore no such memories. They must come to appreciate the achievements of democracy through relevant study and reflection, and also to take part in a relatively active political life. Such things do not come naturally or logically any more. The triumph of individualism has meant that traditional politics matters primarily as a career option. Mainstream media, the workplace and politicians themselves everyday confirm the neoliberal precept that you are on your own. Collective political action is portrayed as something from a grainy black and white past. Schools and universities reinforce the point by downgrading those arts and humanities subjects that are not vocationally oriented.

How then do we get back to basics? This itself a requires a form of political memory, the experience or the recognition that allows us to understand that we can see the world differently and that it is possible to make steps in the direction of change. What indeed are the basics? The framework I have been articulating thus far in this book provides for me what I think are the principal political challenges at the beginning of the 21th century. In finishing, then, let us look at what these are, and what kinds of potentials they contain.

1) *The power of ideas and of collective political action*:

We need to remember the power of political thought and to work to restore that power to institutional politics. Voltaire's quote on the power of ideas: 'No problem can withstand the sustained assault of thinking' is all very inspirational. However, his phrase, or something like it, today encapsulates an ethos that permeates the kinds of books that aspirant CEOs might read, or something that Bill Gates or Steve Jobs would be expected to say, a neat line directed at the individual that has morphed into a vapid business mantra. This is indicative of profound political and cultural change in our society. Voltaire, we need to remember, was no Wall Street venture capitalist or Silicon Valley entrepreneur. He was a philosopher and political reformer whose thinking was fashioned by the spirit of the Enlightenment. Ideas that have the power to change society thus need to be overtly politicized and articulated as democratic in nature and social in their orientation. They need to be seen as useful insofar as they satisfy a collective social need that can be tackled collectively. It needs to be remembered, moreover, that the main problems that require the power of old and new ideas are not difficult to identify. They are the problems that most people could agree to be the most salient. They are the issues that governments across the world easily recognize, but often are unable to do much about, issues such as poverty, health care, housing, jobs, caring for children and the elderly, the environment, global warming and long-term sustainable development.

A proactive and reformist politics would constitute a form of temporalized democracy where the role and function of time (a recognition of the value of a past, present and future orientation) is made salient, if not preeminent. It would be a temporalized politics in that it recognizes the fact that a postmodern 'post-politics' where everything is relative and contingent has not worked, and has only produced anxiety and volatility. Consequently politics could once again take the lead, as it is historically committed to do. This would mean a revitalized polity that sees the social value in the development of explicit political programmes that organize time and space and society—not in a rigid and instrumentalized sense, but in ways that provide open, inclusive and ongoing solutions to the problems that every society contends with. For example, the effects of the five-year plans of the USSR and China are rightly regarded with horror by anyone who cares to read the histories of collectivization and industrialization in these countries. However, the underlying principle behind communist planning was sound, it sought to shape and anticipate the future and its needs, and this was entirely rational. However, the reality of forced labor, of unrealistic targets and of the systemic concealment of disastrous consequences was the result of totalitarian system locked in competition with a 'free-world' world that was equally as totalitarian in its defense of capital. History thus provides a great deal of experience to draw upon in terms of what planning can achieve, what to plan for, and what to leave to the market instead.

This leads to the crucial point of politics taking the lead. A truly democratic society is one where the needs of capital and a narrow stratum of capitalist, either globalized or localized, correspond to the preeminent needs of the majority. This is an age-old socialist principle, or course, but it does not follow that it should be a rigid principle in the form of the always ominous-sounding 'dictatorship of the proletariat'. We need to adopt a different approach. To temporalize a political process can be to suddenly make it look achievable; that is to say, if the political relationship with capital looks to the past and to the future in the ways in which the economy, global or local, is managed, then opportunities present themselves, opportunities that may not be so apparent if the temporal factor is ignored. The approach need not be complex. For instance, a major lesson from the recent past is that the free market, as the leading force in economy and society, has been catastrophic and, accordingly, that the economy requires more democratic management. Likewise, the recent history of social democracy shows that there are also areas in the economy where the market can work better than government. A fundamental 'law' of a temporalized democracy would be to work within an inherent flexibility—with the proviso being that democratic management, however that may be articulated, is the overriding guiding principle. Looking to the future, a temporalized democracy could be expressed through a *political teleology* where, again with an inherent flexibility, the principle of intentionality is foremost. It would require the polity constantly to ask: what kind of society do we want? and what could be the best way to get there? This would be a democratic control over forms of social and economic temporal production, where risks are projected and minimized as much as possible through the ongoing 'stabilization of change' (Adam and Groves, 2007:41–43). In this, politics would function as Enlightenment reason argued that it must, by taking charge of the future, and creating the future through a society that is fully in control of a temporal present that has been rescued from the uncertainties emanating from the abstract laws of unbridled markets and competition.

2) *Remembering the achievements of democracy*:

The good faith and broad based retrieval of the political memory of a city, or region, or nation would be an immensely powerful tool for social change. Such recuperation would bring into play a process of continual assessment and re-assessment of our history and democracy's achievements in the form of an on-going political project; an intrinsic aspect of institutional and cultural political life that ideally could be normalized as a fully functioning element of a incorporative civil society. This conceivably could be a Habermasian public sphere where new technologies of mediation that the network society provides could be used in ways that run counter to their overwhelmingly instrumentalized usage today. What we have seen in the phenomenon of blogging, indeed, provides an insight into what *might* be possible on both a local and global scale.

Politics in such a public sphere could be made both reflective and projective. The political past can be mined for instances, innovations, practices and patterns that have had positive outcomes for the majority of people, and these can serve as policy options to be projected forward to the future.

The post-war experiment with social democracy provides an example of both the achievements of democracy as well as its limitations. The social democratic project, in all its manifestations and in all its local contexts, is a wealth of historical-political memory that can help societies and cultures draw lessons about possible paths for political action in the present and the future. Its major achievements such as socialized healthcare, free universal education, social safety nets and so on could be reconsidered for extension and improvement as opposed to being considered for selling-off. Such a reflective and projective attitude would be mindful of the trap in which the Third Way of the 1980s was snared. Here, a middle way between laissez faire and state socialism was sought, but it was a form of political theory that developed in a historical context where the momentum of rising neoliberalism was too strong, and where many of the Third Way's leading proponents, such as Tony Blair and Bill Clinton, became preoccupied with the speed imperatives of executive office. This factor, coupled with the ideological stifling of any broad-based support for it, meant that was destined not to progress much beyond rhetoric and wishful thinking. The fate of the Third Way, then, is in itself an historical lesson, a memory of the limits to democratic processes when capitalism is given free rein to follow its own logic. It teaches us that along with the limits to democracy, we always need to scrutinize and understand both the limits to capital and the dangers of capital.

A positive remembering of the actual achievements of democracy—instead of hazy idealizations of it—would predispose us to see it in the way that Habermas saw modernity more broadly, as an 'unfinished project' (1987). This would require us to accept that the Enlightenment essence of reason is still intact, but that the instrumental rationality of technology that Marcuse identified (1964) has deflected modernity from its democracy-oriented course. A temporally aware civil society would conclude that after the disasters of neoliberalism, it is politics, not capitalism that has to take the lead again. Moreover, Marcuse's technological rationality, which is always necessary to some degree, needs to be itself deflected by another form of rationality, what Habermas called a 'communicative rationality' wherein the communicative process is oriented towards the social and political instead of, as is currently the case, towards the economic in the profit driven search for efficiencies in society's productive processes. The potential for successful 'communicative rationality' within ICTs is tremendous—but only if people, and not 'market forces', are in control of the processes of technological innovation, development and deployment. To paraphrase Habermas, modernity in a temporalized context, would thus be 'always unfinished' and would develop as part of the longue duree of historical becoming.

3) *Temporal sovereignty as a democratic right:*

The most important idea I have tried to promote is that we must look at temporality in a new way. With the rise of the networked society and neoliberal globalization it is socially and politically necessary to move beyond seeing time as something expressed by the clock, something abstract and uncontrollable that we must synchronize with. The issue is doubly important in the context of network time within the Second Empire of Speed, where social acceleration is now creating myriad social, economic and cultural pathologies.

Time is not 'out there' in the form of some backdrop to the universe, as in the Newtonian sense that has dominated our social, cultural and economic thinking. Time is social; we create time, or what I've argued to be contextually derived 'timescapes', whether we are aware of it or not. It is a process that historically has created the disjuncture we feel, intuitively at least, between social times and the temporalities of the clock and network, one which led St. Augustine of Hippo, writing in the 5th Century, to ask: 'What, then, is time? If no one ask of me I know; if I wish to explain to him who asks, I know not.' By seeing time as social, we can see it as non-abstract and as rooted in concrete processes of everyday life. By seeing time in such a fashion, we would go some way to understanding St. Augustine's dilemma.

Time and space are deeply connected processes. I argued, following Lefebvre (1990) and others, that as time is socially produced, then so too is space. Historically it has been less trouble for us to make the cognitive connection that space is more socially 'real', and have imbued the idea of social space with special properties that recognize this. We thus are able to own space conceived of as 'property' in the form of land, or a house, and so forth, which may be owned by an individual. More broadly we conceive of space as a form of territory that 'belongs' to the people or nation that are contained within it. The idea of sovereignty was developed as a *right* over a territorial area of political governance, a right that inhered down to the level of sovereignty of the 'personal space' of those within its spatial jurisdiction. A logical extension of this idea would be the development of a form of *temporal sovereignty* that would inhere the idea of ownership (individually and/or collectively) into social time.

At the meta-scale the temporal sovereignty of the polity would entail the protection of the institutional processes of democracy. Political time, as Chesneaux (2000) reminded us, would thus be an inviolable time, a sacrosanct temporal relation that could not be sacrificed or rushed for the sake of narrow economic or political expediency. This would constitute a political (democratic) control over time that would recognize that speed has its place and is a requirement that may be accommodated in times of emergency, or when there is a genuine need for rapid action. But it would also be the recognition that many of society's problems *need time* for reflection and deliberation—times that would vary from issue to issue and from problem to problem. A temporalized democracy would thus be equipped to deal with the pathologies of speed that beset society today, and which we have already discussed in some detail. Instead of appearing as unconnected phenomena as they currently do, these pathologies would have a locus, and there would be political recourse to solutions through arguing that these are temporal problems with temporal solutions.

The idea of temporal sovereignty is not so far-fetched. I noted in Chapter Three that in 1999 UN Secretary General Kofi Annan argued (in the context of the expanding network society) that information, and access to it, should be considered a human right. The case for temporal rights is surely just as arguable. What is needed is an acceptance that there is a very deep problem in our relationship with time under a free-market capitalism. This recognition would form the basis for the emergence of the necessary political will to begin to address issues of time in economy, in society and, vitally, in the polity itself.

Temporal rights and temporal sovereignty would feed directly into democratic control over the forms and pace of temporal production in society. If the issue of temporality were made more salient, then the blanket acceleration that we experience under neoliberal globalization would rightly be viewed as illogical, and as ultimately inefficient and wholly unsustainable. We produce time socially and so we need to 'make' it as well as 'save' it. If either making or saving time is considered appropriate to whatever issue confronts us, then resources, temporal and/or technological, should be made available.

These are questions of control and technology that raise the question of the network society itself. I have shown that the networked economy is the primary cause and consequence of social acceleration today, but it does not have to be so. To what extent is it be possible to 'control' the network society? It is neither possible (or desirable) to uninvent the network society or constrain it with crippling state controls. Any such attempts would indicate that we have not remembered or learned the lessons of authoritarian 'democracies' and the disasters they have created. The network society has much potential for the promotion of an affirmative communicative rationality, one where cultures and societies may be positively enhanced through democratic structures that are intensely networked, but in ways that subordinate the free market to social needs. To realize such potential the network society needs to be socialized, brought under democratic control at both the local and global levels (and at the levels in between) instead of being left to develop through the vagaries of competition. A democratically underpinned network society would still create contextual 'timescapes', but the contexts themselves would be structurally oriented toward social needs and desires. Possible the greatest technological prize it that the logic of computing itself would become humanized instead of instrumentalized because people themselves would control it, and not abstract laws. This would be a far more faithful rendering of the dreams of Alan Turing, J.C.R. Licklider and Norbert Weiner—none of whom ever remotely considered a pre-eminent role for market forces in their theorizing.

It is my hope that what has been argued in this concluding section does not read as wishful thinking or, a worse fate, a theory of everything that becomes a theory of nothing. Critical theory and critical thinking are at the core of my work, but this is because we are in the midst of a historical phase, one that is conditioned by what A.C. Grayling (2008) called the 'cancer of the contemporary intellect, postmodernism'—and so critical theory and critical thinking need to be remembered, revived and redeployed to render objective conditions more comprehensible. The first steps are inevitably speculative and tentative

The impetus for the work comes also from the belief that humanity in the 21st century needs something better if it is to see the 22nd in reasonable shape. Today there is no alternative to capitalism, not even anything promising on the horizon. However, through the use of the temporal framework it is possible to provide insights into why the world is as it is today, and how we might begin to change it and direct it to a more positive future. In many ways this is an old message, one that continues to place faith in the power of reason, in the initial impulse of the Enlightenment, and in the potential for democracy to create a better world no matter how impossible it may currently seem, or how lost the cause may today appear. I have tried to articulate a set of provisional alternatives and enact through their essential non-radicality, visions of other ways of being and other ways of seeing. A much wider recognition of the fact that neoliberalism, the free market, and open-ended social acceleration are more a problem than a solution, may be the basis for the building of a global network society that is more democratic and more oriented towards managing the system through networks at the local and the global level. But this in itself would not be the solution. What I've projected thus far is *a politics of reformism.* A more socially- and environmentally-friendly capitalism will not solve capitalism's contradictions. They will always rise to the surface in the form of crises at some point in time and space, sooner or later. For a temporalized democracy to work fully and towards its open potentiality, capitalism has to be got rid of. There can be no such thing as a permanently viable model of capitalism. I see the present phase, Debray's 'videosphere', as much less durable and profound in its effects as the 'graphosphere' because it is based upon rapidly evolving technological change. And so we are living through an interregnum, or breathing space: a time where, despite continual change and distraction, we need to imagine viable alternatives to the rule of capital and think about how these may be articulated, developed and implemented. Control over the spaces and times of the network would be the first step towards a new global imaginary. In the current void that represents other ways of thinking about how economic, political and temporal life might be ordered, it stands as the only way forward.

REFERENCES

Adam, Barbara (1998) *Timescapes of Modernity: The Environment and Invisible Hazards.* London: Routledge.

—— (2004) *Time* Cambridge: Polity.

Adam, Barbara & Groves, Chris (2007) *Future Matters* Leiden: Brill Publishers.

Adams, Henry (1918) 'A Law of Acceleration (1904)' *Henry Adams: The Education of Henry Adams* http://www.bartleby.com/159/34.html.

Adorno, Theodor & Horkheimer, Max (1986) *The Dialectic of the Enlightenment* London: Verso.

Agger, Ben (1992) *The Discourse of Domination* Evanston: Northwest University Press.

Ainsworth, W.H. (1834) *Rookwood* Philadelphia: Carey, Lea & Blanchard.

American Bar Association (2006) See ABA Recommendations Report at: http://www.abanet.org/op/signingstatements/aba_final_signing_statements_recommendation-report_7-24-06.pdf.

Angell, Marcia (2004) *The Truth About the Drug Companies: How They Deceive Us and What to Do About It* New York: Random House.

Anderson, Perry (2004) 'The River of Time' *New Left Review* 26 March–April.

Annan, Kofi (1999) 'ITU Opening Ceremony' http://www.itu.int/telecom-wt99/press_service/information_for_the_press/press_kit/speeches/annan_ceremony.html.

Apple.com (2006) http://www.apple.com/itunes/1billion/

—— (2007) http://www.apple.com/pr/library/2007/07/31itunes.html.

Bachelard, Gaston (1996) *The Poetics of Space* New York: Orion Press.

Barber, Benjamin (2007) *Consumed* New York: W.W. Norton.

Barnett, Ronald (1997) *Higher Education: A Critical Business.* Buckingham: Open University Press.

Barney, Darin (2004) *The Network Society* Oxford: Polity.

Baudrillard, Jean (1988) 'Symbolic Exchange and Death' in *Selected Writings Mark* Poster (ed) Cambridge: Polity Press.

Bauman, Zygmunt (1998) *Globalization: The Human Consequences*, Cambridge: Polity Press.

BBC NewsOnline (2005) 'Iran jails blogger for 14 years' http://news.bbc.co.uk/2/hi/technology/4292399.stm.

Beams, Nick (2004) 'US deficit to hit half a trillion dollars' Worldwide Socialist Web http://www.wsws.org/articles/2004/feb2004/defi-f04.shtml.

Beck, Ulrich, & Beck-Gernsheim, Elizabeth (2002) *Individualization: Institutionalized Individualism and its Social and Political Consequences*, Thousand Oaks, Calif: Sage.

Beck, Ulrich (2004) 'Cosmopolitical Realism: On the Distinction Between Cosmopolitanism and Realism in Philosophy and the Social Sciences' *Global Networks* 4, 2 pp. 131–156.

Bell, Daniel (1973) *The Coming of the Post-Industrial Society* New York: Basic Books.

Bell, E.T. (1953) *Men of Mathematics* Harmondsworth: Penguin.

Bello, Walden (2005) *Dilemmas of Domination* New York: Henry Holt.

Benedict, Michael (1993) *Cyberspace: First Steps* Cambridge, Mass.: MIT Press.

Benjamin, Walter (1936) 'The Storyteller: Reflections on the Works of Nikolai Leskov' http://www.slought.org/files/downloads/events/SF_1331-Benjamin.pdf.

Berger, John (1998) 'Against the Great Defeat of the World' *Race and Class* Vol. 40, Number 2/3 p. 1.

Bergson, Henri (1946) *The Creative Mind: An Introduction to Metaphysics* New York: Kensington Published.

—— (2001) *Time and Free Will: An Essay on the Immediate Data of Consciousness* (c. 1913), reprinted Mineola, NY: Dover Publications, Inc.

Berman, Marshall (1981) *All That is Solid Melts into Air: the Experience of Modernity* London: Verso.

Bertman, Stephen (1999) *Hyperculture: The Human Cost of Speed* Westport, CT: Praeger.

Bingham, Nick & Thrift, Nigel (2000) 'Some new instructions for travellers: the geography of Bruno Latour and Michel Serres' *Thinking Space* Mike Crang & Nigel Thrift (eds) London and New York: Routledge.

Blechman, Rob (2005) *Media Ecology* http://www.media-ecology.org/.

Bobbio, Norberto (1987) *The Future of Democracy: A Defence Of The Rules Of The Game* Minneapolis: University of Minnesota Press.

Brautigan, Richard (1967) 'All Watched over by the Machines of Loving Grace' http://www.cs.unca.edu/~edmiston/poems/grace.html.

Brenner, Robert (2003) 'Towards the Precipice' *London Review of Books* vol. 25 number 3, February.

Bronner, Stephen, Eric (1995) 'The Great Divide: The Enlightenment and its Critics' *New Politics*, Vol. 5, no. 3 No. 19, Summer.

BNET (2007) 'CDC Games Achieves Rapid Growth in Registered Users for Its Open Beta Test of Special Force' www.cdcgames.net/cdcgames/investor.html.

Bunting, Madeline (2005) *Willing Slaves: How the Overwork Culture is Ruling Our Lives* New York: Harper Collins.

Callinicos, Alex (1991) *The Revenge of History: Marxism and the East European Revolutions* Cambridge: Polity Press.

Castells, Manuel (1997) *Power of Identity* Vol 2 of *The Information Age: Economy, Society and Culture* Oxford: Blackwell.

—— (1996) *The Rise of the Network Society* Vol 1 *The Information Age: Economy, Society and Culture* Oxford: Blackwell.

—— (1999) 'Information Technology, Globalization and Social Development' UNRISD Discussion Paper No. 114, September. http://www.unrisd.org/80256B3C005BCCF9/(httpAuxPages)/F270E0C066F3DE7780256B67005B728C/$file/dp114.pdf.

—— (2001) The Internet Galaxy Oxford: Oxford University Press.

Catan, Thomas (2007) 'Late-running nation told to wake up and start living in English time' http://www.timesonline.co.uk/tol/news/world/us_and_americas/article1449469.ece

Cheng, Jacqui (2008) 'Blogosphere growth slowing considerably' Ars Technica http://arstechnica.com/news.ars/post/20070405-blogosphere-growth-slowing-considerably.html.

Chesneaux, Jean (2000) 'Speed and Democracy' *Social Science Information*, Vol. 39, No. 3, 407–420.

Chmielewski, Dawn (2004) 'Sony to launch iPod counterattack' http://www.ecommercetimes.com/story/34945.html.

CIRAC (2005) 'Creative Industries' http://www.creativeindustries.qut.com.

Cockburn, Alexander & St.Clair, Jeffrey (2004) *Imperial Crusades* London: Verso.

Copeland, Douglas (1991) *Generation X: Tales for an Accelerated Culture*. New York: St. Martin's Press.

Daley, James (2005) 'This boot camp can make kids cash wise' *The Independent Online* http://www.independent.co.uk/money/invest-save/this-boot-camp-can-make-kids-cash-wise-484757.html.

Davies, Norman (1997) *Europe: A History* London: Pimlico.

Davis, Mike (2005) 'The Predators of New Orleans' *Le Monde Diplomatique* http://mondediplo.com/2005/10/02katrina.

Davies, Mike and Monk, Daniel Bertrand (2008) *Evil Paradise: Dreamworlds of Neoliberalism* New York: The New Press.
Dawson, Ross (2003) *Living Networks* London: FTPrentice Hall.
De Zengotita, Thomas (2005) *Mediated: How the Media Shape Your World* London: Bloomsbury.
Debray, Regis (1973) 'Time and Politics' in *Prison Writings* New York: Random House.
—— (2007) 'Socialism: A Life-Cycle' *New Left Review* 46 July–August.
Debord, Guy (1983) *The Society of the Spectacle*, Michigan: Black and Red Press.
Delanty, Gerard (2001) *Challenging Knowledge: The University in the Knowledge Society* Buckingham: Open University Press, 2001.
Deleuze, Gilles and Guattari, Felix (1994) *What Is Philosophy?* New York: Columbia University Press.
Deleuze, Gilles (1997) 'Intellectuals and Power: A Conversation Between Gilles Deleuze and Michel Foucault' http://info.interactivist.net/node/1567.
—— (1997a) 'Intellectuals and Power' in D. Bouchard (ed) *Language, Counter-Memory, Practice*, Ithaca: Cornell University Press.
Derrida, Jacques (1993) *Specters of Marx, the State of the Debt, the Work of Mourning, & the New International*, translated by Peggy Kamuf, Routledge.
Dunant, Sarah & Porter, Roy (1996) (eds) *The Age of Anxiety* London: Virago.
Eagleton, Terry (1991) *Ideology*. London: Verso.
EGS (2008) European Graduate School http://www.egs.edu/.
Edwards, Paul N. (1995) *The Closed World: Computers and the Politics of Discourse in Cold War America* Cambridge, MA: MIT Press.
Elias, Norbert (1992) *Time: An Essay* Oxford: Blackwell.
Elliot, Larry & Boseley, Sarah (2007) 'Brown calls on Google to help world's poor' *The Guardian* (10th December).
Eriksen, T.H. (2000) *Tyranny of the Moment: Fast and Slow Time in the Information Age*. London: Pluto.
Ewen, Stuart (1976) *Captains of Consciousness: Advertising and the Social Roots of the Consumer Culture*, McGraw-Hill Book Company.
Fairbanks, Amanda M. (2008) 'The Year of the Political Blogger' *New York Times* online, 22nd August http://www.nytimes.com/2008/08/24/fashion/24blog.html.
Fasolt, Constantin (2004) Review of Gerhard Dohrn-van Rossum's *Die Geschichte der Stunde: Uhren und moderne Zeitordnungen*, München: Carl Hanser, 1992 *Bryn Mawr Classical Review* 06:19.
Feenberg, Andrew (2004) *Questioning Technology* London: Routledge.
Flood, A. (1996) 'Report on the First Intercontinental gathering for Humanity Against Neoliberalism' http://flag.blackened.net/revolt/andrew/encounter1_report.html.
Flora, Peter (1981) *The Development of Welfare States in Europe and America* Edited by Peter Flora and Arnold J. Heidenheimer New Brunswick, U.S.A: Transaction Books.
Forex.com (2008) 'Foreign Exchange Markets' http://www.forex.com/08-06-30.html.
Foster, W., Seymour E.G. (2000) 'The Diffusion of the Internet in China' *Centre for International Security and Cooperation* (CISAC), Stanford University, November. http://mosaic.unomaha.edu/china_2000.pdf.
Foster, Hal (1983) *The Anti-Aesthetic: Essays on Postmodern Culture* Washington: Bay Press.
Foucault, Michel (1970) *The Order of Things* New York: Pantheon.
Frank, Thomas (2000) *One Market Under* God London: Secker & Warburg.
Frankel, Boris (2001) *When The Boat Comes In* Sydney: Pluto Press.
Fraser, J.T. (2003) 'Time Felt, Time Understood' *KronoScope*, Volume 3, Number 1, 2003, pp. 15–26(12).
Freud, Sigmund (1979) *On Psychopathology: Inhibitions, Symptoms and Anxiety, and Other Works* Harmondsworth: Penguin.

Frow, John (1995) *Cultural Studies and Cultural Value* Oxford: Clarendon Press.
Fukuyama, Francis (1992) *The End of History and the Last Man* New York: Free Press.
Furedi, Frank (2004) *Therapy Culture: Cultivating Vulnerability in an Uncertain Age* London: Routledge.
Gallagher, Michael (2008) 'Protecting the Video Game Industry's Value Creation and Growth' ww.perkinscoie.com/. . ./Presentation/File/Game%20Industry%20Round table%20-%20ESA.pps.
Galtung, Johann (2005) 'On the coming Decline and Fall of the US Empire' *Transnational.org.* http://www.transnational.org/forum/meet/2004/Galtung_USempireFall.html.
Gare, Arran (1996) *Nihilism Inc.* Sydney: Eco-Logical Press.
Garton-Ash, Timothy (2005) 'Soldiers of the Hidden Imam' *New York Review of Books* Vol. 52 Number 17.
Gates, Bill (2000) *Business @ The Speed of Thought* New York: Warner Business Books.
Geertz, Clifford (1993) The Interpretation of Cultures New York: Basic Books.
Giddens, Anthony (1990) The Consequences of Modernity. Cambridge: Polity Press.
—— (1999) Runaway World London: Profile Books.
Gies, Frances & Gies, Joseph (1995) *Cathedral, Forge, and Waterwheel: Technology and Invention in the Middle Ages* New York: HarperCollins.
Glasner, Joanna (2005) 'P2P Fuels Global Bandwidth Binge' *Wired* http://www.wired.com/news/business/0,1367,67202,00.html.
Gleick, J. (1999) *Faster: The Acceleration of Just About Everything*, New York: Abacus.
Glyn, Andrew (2005) 'Imbalances of the Global Economy' *New Left Review* 34, July–August.
Gohring, Nancy (2006) 'Cell phone Sales Topped 800 Million in 2005' http://www.infoworld.com/article/06/01/27/74861_HNmobilephonesales_1.html.
Goldberg, Ken (2001) *The Robot in the Garden: Telerobotics and Telepistemology in the Age of the Internet* MIT Press.
Goldman, Robert (2004) 'Placeless Places' *Representations of Global Capital* http://www.lclark.edu/~soan370/global/nortelsign33.html.
Goldman, Robert, Papson, Stephen & Kersey, Noah (2004a) 'Speed: Through, Across and In the Landscape of Capital' *Fast Capitalism* 1:1.
Grayling, Anthony (2008) 'Origin of the Specious' The New Humanist Volume 123 Issue 5 September/October http://newhumanist.org.uk/1856.
Greenpeace (2004) 'Genetically Engineered Fish Under Development' http://www.greenpeaceusa.org/ge/fish_varieties_text.htm.
Grosz, Elizabeth (1998) 'Thinking the New: Of Future Yet Unthought' Symploke 6.1 38–55.
Habermas, Jurgen (1987) The Theory of Communicative Action Boston: Beacon Press.
Haraway, Donna (1991) 'A Cyborg Manifesto: Science, Technology and Socialist-Feminism in the Late Twentieth Century,' in *Simians, Cyborgs and Women: The Reinvention of Nature* New York: Routledge.
Hardt, Michael and Negri, Antonio (2000) Empire, Harvard University Press.
Hardt, Michael (2002) 'Porto Alegre: Today's Bandung?' New Left Review 14, March–April. pp. 112–118.
—— (2005) Multitude: War and Democracy in the Age of Empire New York: Penguin Press.
Harvey, David (1982) *The Limits to Capital*, Cambridge: Blackwell.
—— (1989) *The Condition of Postmodernity*, Cambridge: Blackwell.
—— (2004) *The New Imperialism* Oxford: Oxford University Press.
Hassan, Robert (2003) 'The MIT Media Lab: techno dream factory or alienation as a way of life?' *Media, Culture and Society* Vol. 25 Number 1 pp. 87–107.
—— (2004) *Media, Politics and the Network Society* Maidenhead: Open University Press.

—— (2005) 'Liquid Time and Space' *Southern Review* Volume 38 Number 2.
Hassan, Robert and Purser, Ronald (2007) *24/7: Time and Temporality in the Network Society* Stanford, California: Stanford University Press.
Hassan, Robert (2008) *The Information Society* Oxford: Polity.
Hayek, Friedrich von (1978) *The Constitution of Liberty* Chicago: the University of Chicago Press.
Heim, Michael (1993) *The Metaphysics of Virtual Reality* New York: Oxford University Press.
Held, David (1987) *Models of Democracy*, Stanford: Stanford University Press.
Hesseldahl, Arik (2004) 'The wont-be iPod killer' *Forbes.com* http://www.forbes.com/technology/feeds/wireless/2004/06/14/wireless01087222502107-20040614-004600.html.
Hobsbawm, Eric (1996) *The Age of Capital 1848–1875* London: Weidenfeld & Nicholson.
Hochschild, Arlie (1997) *The Time Bind* New York: Henry Holt.
Hopkins, Terence K. & Wallerstein, Immanuel (1996) *The Age of Transition: Trajectory of the World-System* 194–2025 London & New Jersey: Zed Books.
Hörning, Karl H., Ahrens, Daniela, and Gerhard, Annette (1999) 'Do Technologies Have Time?' *Time & Society*, Vol. 8(2): 293–308.
Husserl, Edmund (1964) The Phenomenology of Internal Time Consciousness (1905–10), trans. S. Churchill, The Hague: Martinus Nijhoff.
Hutton, Will & Giddens, Anthony (2001*) On the Edge: Living With Global Capitalism*, London: Verso.
Hyman, Anthony (1982) *Charles Babbage: Pioneer of the Computer* Oxford: Oxford University Press.
Isserman, Maurice (2002) 'Ben Franklin and the Gulf Stream' http://www.studyofplace.com/ModuleContent/OceanCurrentsReading.html.
Jameson, Fredric (1992) *Late Marxism: Adorno, Or, The Persistence of the Dialectic* London: Verso.
—— (1996) 'Five Theses on Actually Existing Marxism' *Monthly Review* Volume 47, Number 11, April.
Jessop, Bob (2007) 'What follows neo-liberalism? The deepening contradictions of US Domination and the Struggle for a New Global Order'. http://eprints.lancs.ac.uk/1005/.
Jones, Gerard (2002) *Killing Monsters* New York: Basic Books.
Joseph, Emma (2005) *Analysis* BBC World Service 'Video Games' Broadcast 5th June.
Judt, Tony (2004) 'Dreams of Empire' *New York Review of Books* Vol. 51, Number 17, November 4. http://www.nybooks.com/articles/article-preview?article_id=17518.
—— (2007) 'The Wrecking Ball of Innovation' *New York Review of Books Online* Volume 54, Number 19 December 6 http://www.nybooks.com/articles/20853.
Jung, Carl, Gustav (1980) *Memories, Dreams, Reflections*. New York: Random House.
Kalyvas, Andreas (2003) 'Feet of Clay? Reflections on Hardt and Negri's *Empire*' *Constellations* Volume 10, No 2, pp. 264–279.
Kelly, Kevin (1998) 'The Computational Metaphor' *Whole Earth* (Winter) http://www.wholeearth.com/ArticleBin/201.html.
Kenyon, Susan (2008) 'The Prevalence of Multitasking and the Influence of Internet Use' *Time & Society* Vol. 17 Number 2/3.
Kenward, Rod (2005) *La Vie en Bleu* London: Allen Lane.
Keynes, John Maynard (1926) 'The End of Laissez Faire' http://www.panarchy.org/keynes/laissezfaire.1926.html.
KCNN (2008) Knight Citizen News Network 'Citizen Journalism: Blogging: The Invasion of the Horde' http://www.kcnn.org/research/citizen_journalism_blogging.
Klein, Naomi (2000) No Logo: *Taking Aim At The Brand Bullies*. London: Flamingo.
—— (2002) *Fences and Windows*, London: Flamingo.
—— (2007) *The Shock Doctrine* London: Penguin Allen Lane.

Klein, Olivier (2004) 'Social Perception of Time, Distance and High-Speed Transportation' *Time & Society* Vol. 13 No. 2/3. pp. 245–263.
Kolko, Joyce (1988) *Restructuring the World Economy* New York: Pantheon.
Kuipers, Dean (2000) I Am a Bullet: Scenes from an Accelerating Culture New York: Crown Publishing.
Lakoff, George and Johnson, Mark (1980) *The Metaphors We Live By* Chicago: University of Chicago Press.
Lanchester, John (2008) 'Cityphilia' *London Review of Books*, January 3 http://www.lrb.co.uk/v30/n01/lanc01_.html.
Landes, David (1983) *Revolution in Time: Clocks and the Making of the Modern World* Cambridge MA: The Belknap Press.
Lash, Scott (2002) *Critique of Information* London: Sage Publications.
Latour, Bruno (1997) 'Trains of Thought: Piaget, Formalism, and the Fifth Dimension' *Common Knowledge*, 6.3: pp. 170–191.
—— (2002) 'Morality and Technology' *Theory, Culture & Society*, Vol. 19, No. 5–6: 247–260.
Le Goff, Jacques (1980) *Time, Work and Culture in the Middle Ages* Chicago: University of Chicago Press.
Leal, P. (2000) 'Participation, communication and technology in the age of the global market' *Forest, Tree and People* (Newsletter) Number 40/41 http://www-trees.slu.se/newsl/40/40leal.pdf.
Lebkowsky, Jon (2004) 'The Future of Affinity: Living Networks with Social Software' Presented to the CenTex Chapter of the World Future Society, *November 18, 2004* http://www.weblogsky.com/archives/000276.html.
Lechner, Frank J. and John Boli (eds.) (2007) *The Globalization Reader* Oxford: Blackwell.
Lee, Heejin & Liebenau, Jonathan (2000) 'Time and the Internet at the Turn of the Millennium', *Time & Society*, Vol. 9, No. 1, pp. 43–56, 2000.
Lefebvre, Henri (1991) *The Production of Space* Oxford: Basil Blackwell.
Lehmann, Nicholas (1998) 'A Fool's Goal' *Forbes.com* http://www.forbes.com/asap/1998/1130/281.html 1998.
Lessig, Lawrence (1999) *Code and Other Laws of Cyberspace* New York: Basic Books.
—— (2005) 'The People Own Ideas!' *TechnologyReview.com* http://www.technologyreview.com/articles/05/06/issue/feature_people.asp?p=0.
Licklider, J.C.R. (1960) 'Man-Computer Symbiosis,' in *IRE Transactions on Human Factors in Electronics*, Vol HFE-1, March, pp. 4–11.
Lipset, Seymour Martin (1960) *Political Man: The Social Bases of Politics* New York: Doubleday.
Livingstone, David & Withers, Charles (1999) *Geography and Enlightenment* Chicago: University of Chicago Press.
Lowe, Janet (1998) *Jack Welch Speaks* New York: John Wiley & Sons.
Loy, David R. (2000) 'The Spiritual Origins of the West' in *International Philosophical Quarterly* Vol. XL No. 2, pp. 215–235.
Lunenberg, Peter (2000) *The Digital Dialectic: New Essays on New Media* Boston: MIT Press.
Lyon, David (2002) *The Surveillance Society*, Buckingham: Open University Press.
Lyotard, Jean-Françoise (1979) *The Postmodern Condition: A report on knowledge* Manchester University Press.
Manovich, Lev (2001) 'What is New Media?' in *The Language of New Media* Cambridge, Mass: MIT Press.
Marcuse, Herbert (1964/1991) *One Dimensional Man* Boston: Beacon Press.
Marginson, Simon & Considine, Mark (2000) *The Enterprise University: Power, Governance and Reinvention in Australia*, Cambridge: Cambridge University Press.
Martin, David (2007) 'Spilt Religion' *Times Literary Supplement*, August 10.

Marx, Karl & Engels, Friedrich (1975) 'The Manifesto of the Communist Party' in *Selected Works* Moscow: Progress Press.
Marx, Karl (1973) *Grundrisse: Foundations of the Critique of Political Economy* Harmondsworth, Middlesex: Penguin Books.
Mattelart, Armand (1991) *Advertising International* London: Routledge.
—— (1995) 'Les nouveaux scenarios de la communication mondiale' *Le Monde Diplomatique*, August, pp. 24–25.
McChesney, Robert & Herman, Edward (1997) *The Global Media: The New Missionaries of Corporate Capitalism* London: Cassell.
McChesney, Robert (2004) *The Problem of the Media* New York: Monthly Review Press.
McCarthy, Rory (2003) 'Salam's Story' *The Guardian* May 30th. http://www.guardian.co.uk/g2/story/0,3604,966768,00.html.
McHoul, Alex (1998) 'Cybernetymology and Ethics' http://www.iath.virginia.edu/pmc/text-only/issue.998/9.1mchoul.txt.
McKenzie, Donald & Wajcman, Judy (eds) *The Social Shaping of Technology*, Buckingham: Open University Press.
McKinnon, Rebecca (2005) RConversation Weblog (http://rconversation.blogs.com/rconversation/2005/06/confirmed_all_t.html.
McLuhan, Marshal (1964) *Understanding Media* London: Sphere Books.
—— (1988) *Laws of Media: The New Science* Toronto; Buffalo: University of Toronto Press.
McNally, David (1998) 'Globalization on Trial' *Monthly Review* Vol. 50 Number 4.
Michels, Robert (1966) *Political Parties* New York: The Free Press.
Mills, C. Wright (1970) *The Power Elite* New York: Oxford University Press.
MIT News (2001) 'NSF awards $13.75M to MIT Media Lab to create Center for Bits and Atoms' http://web.mit.edu/newsoffice/2001/bitsandatoms.html.
Misztal, Barbara A. (2003) *Theories of Social Remembering* Maidenhead: Open University Press.
Monbiot, George (2000) *The Captive State* Basingstoke: Macmillan.
—— (2007) 'Hurray! We're Going Backwards!' Monbiot.com http://www.monbiot.com/archives/2007/12/17/hurray-were-going-backwards/.
Moody, George (1997) 'The Greatest OS That Never Was' *Wired* Issue 5.08 http://www.wired.com/wired/archive/5.08/linux.html?topic=&topic_set=.
Moore, Carla (2006) 'Two Billion GSM Phone Users Worldwide' http://www.dmeurope.com/default.asp?ArticleID=16019.
Mumford, Lewis (1934/1967) *Technics and Civilization*, London: Routledge & Kagan Phillip.
Neale, Martha (1984) *The Development of Logic* Oxford; Oxford University Press.
Negroponte, Nicholas (1995) *Being Digital*. New York: Vintage.
News.nanoapex (2002) http://news.nanoapex.com/modules.php?name=News&file=article&sid=2904.
Nolan, James (1998) *The Therapeutic State*, New York: New York University Press.
Norris, Christopher (1992) *Uncritical Theory: Postmodernism, Intellectuals, and the Gulf War*. Amherst: The University of Massachusetts Press.
North, Douglass (2004) *Understanding the Process of Economic Change* Princeton: Princeton University Press.
Nowotny, Helga (1989) *Eigenzeit* Frankfurt am Main: Suhrkamp Verlag.
—— (1994) *Time: The Modern and Postmodern Experience* Cambridge: Polity Press.
Packer, George (2005) *The Assassins' Gate: America in Iraq* New York: Farrar, Straus and Giroux.
Padgham, Jane (2007) 'Bankruptcies hit record level as consumer debt reaches crisis point' *The Independent* online, 7th February http://www.independent.co.uk/news/business/news/bankruptcies-hit-record-level-as-consumer-debt-reaches-crisis-point-434872.html.

Patterson, T. (1997) *Grand Expectations: The United States 1945–1973* Oxford: Oxford University Press.

Pain, Julien (2005) 'Bloggers: The New Heralds of Free Expression' http://www.rsf.org/article.php3?id_article=14998.

Pask, Gordon (1982) *Micro Man: Computers and the Evolution of Consciousness* London: Macmillan.

Pew Internet (2005) 'A Decade of Adoption' http://www.pewinternet.org/PPF/r/148/report_display.asp.

—— (2008) 'Search Engine Use' http://www.pewinternet.org/PPF/r/258/report_display.asp.

Pierson, Chris (2001) 'Globalization and the End of Social Democracy' *Working Documents in the Study of European Governance* Number 9 (May).

Pierson, Paul (2004) *Politics in Time* Princeton, NJ: Princeton University Press.

Pieterse, Jan (2004) 'Neoliberal Empire' Theory, Culture & Society Vol 21 (3).

Polaski, Sandra (2004) 'Job Anxiety is Real and it's Global' Carnegie Endowment Policy Brief Number 30, May. http://www.carnegieendowment.org/files/Policybrief30.pdf.

Polanyi, Karl (1944/1957) *The Great Transformation* Boston: Beacon Hill Press.

Postman, Neil (1993) *Technopoly: The Surrender of Culture to Technology* New York: Vintage.

Prendergast, Christopher (2003) 'Codeword Modernity' New Left Review' 24 November–December.

Provigil (2008) www.provigil.com.

Putnam, Robert (2000) Bowling Alone: *The Collapse and Revival of American Community* New York: Simon and Schuster.

Quarterman, J.S. (2002) 'Monoculture Considered Harmful' *First Monday* Issue 2, February http://firstmonday.org/issues/issue7_2/quarterman/index.html.

Ransford, Marc (2005) 'Average person spends more time using media than anything else' Press Release Ball State University http://www.bsu.edu/news/article/0,1370,7273-850-36658,00.html.

Rao, Madanmohan & Mendoza Lunita. *Asia Unplugged: The Wireless and Mobile Media Boom in the Asia-Pacific* London: Response Books.

Reid, Herbert G. (1973) 'American Social Science in the Politics of Time and the Crisis of Technocorporate Society: Toward a Critical Phenomenology' *Politics and Society* Winter, 3 (2).

Reporters Without Borders (RWB) (2008) 'IOC accepts organized online censorship' http://www.rsf.org/article.php3?id_article=27988.

Reuters (2004) 'Ford to Triple China Output to Chase GM' http://www-rohan.sdsu.edu/~rgibson/FordandGMChina.html.

—— (2008) 'Recovery may take longer than usual: Greenspan' http://www.reuters.com/article/businessNews/idUSDXB00015620080225.

Rheingold, Howard (1993) *The Virtual Community* Reading, Mass: Addison-Wesley.

Rifkin, Jeremy (1987) *Time Wars* New York: Henry Holt.

—— (2001) *The Age of Access* London: Penguin.

Rippin, Hannah (2005) 'The Mobile Phone in Everyday Life' *Fast Capitalism* 1.1 http://www.uta.edu/huma/agger/fastcapitalism/1_1/rippin.html.

Roberts, Paul (2004) *The End of Oil* Boston, MA. Houghton Mifflin.

Robertson, Roland (1992), *Globalization: Social Theory and Global Culture*, Sage Publications, London.

Robins, Kevin (1996) *Into the Image: Culture and Politics in the Field of Vision* London: Routledge.

Robins, Kevin and Webster, Frank. (1999) *Times of the Technoculture: From the Information Society to the Virtual Life* London: Routledge.

Robinson, William I. (1996) *Promoting Polyarchy—Globalization, US intervention and hegemony* Cambridge University Press.

Rosa, Hartmut (2003) 'Social Acceleration' *Constellations* 10, Number 1.

Rosa, Hartmut and Scheuerman, William (2008) (eds) High-Speed Society: Social Acceleration, Power and Modernity PA: Penn State University Press.
Rosen, Jeffrey (2005) *The Naked Crowd: Reclaiming Security and Freedom in an Anxious Age* New York: Random House.
Rossetto, Lois (1998) *Wired* 6.01 January http://www.wired.com/wired/archive/6.01/toc .html.
Roszak, Theodore (1986) *The Cult of Information* The University of California Press.
Roth, Guenther (2003) 'The Near-Death of Liberal Capitalism: Perceptions from Weber to the Polanyi Brothers' *Politics and Society* Vol.31 Number 2.
Roush, Wade (2007) '*Upheaval at MERL: Mitsubishi Electric Breaks Up Famous Computer Science Lab*' http://www.xconomy.com/2007/07/31/upheaval-at-merl-mitsubishi-electric-breaks-up-famous-computer-science-lab/.
Rucht, Dieter (1999) 'Michel's Iron Law of Oligarchy Reconsidered' *Mobilization* Vol. 4 Number 2.
Russia Today (2008) 'Online game rivalry ends with real life murder' http://www .russiatoday.com/scitech/news/1977.
Sabelis, Ida (2002) *Managers' Times* Amsterdam: Bee's Books.
—— (2004) 'Global Speed: A Time view on Transnationality' *Culture and Organization* Vol.10 (4), pp. 291–301.
Saul, John Ralston (2005) *The Collapse of Globalism and the Reinvention of the World* London: Viking.
Scheuerman, William E. (2002) 'The Economic State of Emergency' *Cardozo Law Review* 21.
—— (2003) 'Global Law in Our High Speed Economy' http://www.isanet.org/noarchive/ scheuerman.html.
—— (2004) Liberal Democracy and the Social Acceleration of Time Johns Hopkins University Press.
Schiller, Dan (1999) *Digital Capitalism: Networking the Global Market System*, Cambridge, Mass: MIT Press.
Schlosser, Eric (2002) *Fast Food Nation* New York: Perennial.
Schor, Juliet (1993) *The Overworked American* New York: Basic Books.
Sennett, Richard (1998*) The Corrosion of Character: The Personal Consequences of Work in the New Capitalism* New York: W.W. Norton and Company.
Servon, Lisa J. (2002) *Bridging the Digital Divide* Oxford: Blackwell.
Shenk, David (1997) Data Smog New York: Abacus.
Showalter, Elaine (1985) *The Female Malady: Women, Madness, and English Culture, 1830–1980.* New York: Pantheon.
Simmel, Georg (1907/2008) 'The Pace of Life and the Money Economy' in High-Speed Society: Social Acceleration, Power and Modernity Hartmut Rosa and William Scheuerman (eds) PA: Penn State University Press.
Silverman, Erik Kline (1992) 'Clifford Geertz: Towards a more "thick" understanding?' in *Reading Material Culture*, Christopher Tilley (ed) Oxford: Blackwell.
Simon, Herbert (1971) *Computers, Communications and the Public Interest* Martin Greenberger, Baltimore, Johns Hopkins University Press.
Simpson, Lorenzo (1995). *Technology, Time and the Conversations of Modernity* NY: Routledge.
Slater, Don (1997) *Consumer Culture and Modernity* Oxford: Polity.
Smith, Adam (1776/1965) *An Inquiry into the Nature and causes of the Wealth of Nations*, New York: The Modern Library.
Sokal, Alan (1996) 'Transgressing the Boundaries: Towards a Transformative Hermeneutics of Quantum Gravity' http://www.physics.nyu.edu/faculty/sokal transgress_v2/ transgress_v2_singlefile.html.
Stallman, Richard (2002) *Free Software, Free Society* Boston: Free Software Foundation.
Steger, Manfred (2001) *Globalism: The New Market Ideology* Lanham: Rowman & Littlefield Publishers.

Stein, Jeremy (2001) 'Reflections on Time, Time-space-compression and Technology in the Nineteenth Century', in Jon May and Nigel Thrift *Timespace*, Jon May & Nigel Thrift (eds) London: Routledge.

Sunstein, Cass (2001) *Republic.com* Princeton, N.J.: Princeton University Press.

—— (2007) *Republic.com 2.0* Princeton N.J.: Princeton University Press.

Tapscott, Don (1996) *The Digital Economy* New York: McGraw-Hill.

Teeple, Gary (1995) *Globalization and the Decline of Social Policy* Toronto: Garamond.

Thompson, E.P. (1993/1967) 'Time, Work-Discipline and Industrial Capitalism' in *Customs in Common* London: Penguin Books.

Thomson, Tony (2005) 'All work and no play in the "virtual sweatshop"' *The Observer* Sunday 13th March.

Thrift, Nigel and Taylor, Michael (1982) *The Geography of Multinationals: Studies in the Spatial Development and Economic Consequences of Multinational Corporations*, London: Croom Helm.

Thrift, Nigel (1996) *Spatial Formations* London: Sage.

—— (2001) 'Chasing Capitalism' *New Political Economy* Vol. 6, No. 3 pp. 375–379.

Thrift, Nigel & May, Jon (2001) *Timespace* London/New York: Routledge.

Tillich, Paul (1980) *The Courage to Be* New Haven: Yale University Press.

Time Magazine (1992) 'How Sam Walton Got Rich' June 15th http://www.time.com/time/time100/builder/profile/walton2.html.

Toffler, Alvin (1970) *Future Shock* London: Pan.

Tofts, Darren (1997) *Memory Trade: A Prehistory of Cyberculture* Sydney: 21C Press.

Tomlinson, John (2007) *The Culture of Speed: The Coming of Immediacy* London: Sage.

Toynbee, Polly (2005) 'The Chasm Between Us' *The Guardian Weekly* September 16–22, p. 13.

Turner, Graham (2008) *The Credit Crunch: Housing Bubbles, Globalisation and the Worldwide Economic Crisis* London: Pluto Press.

UNDP (2005) Report on the World Situation 2005: *The Inequality Predicament* Geneva.

Virilio, Paul (1995) 'Speed and Information Cyberspace Alarm!' *ctheory* August 27th http://www.ctheory.net/text_file.asp?pick=72.

—— (1986) Speed and Politics New York: Semiotext(e).

—— (1997). *Open Sky* (trans. Julie Rose). London: Verso.

—— (2000) 'Interview with John Armitage' http://www.ctheory.net/articles.aspx?id=132.

—— (2008) 'The state of Emergency' in High-Speed Society: Social Acceleration, Power and Modernity Hartmut Rosa and William Scheuerman (eds) Penn State University Press.

Weber, Max (1989) *The Protestant Ethic and the Spirit of Capitalism* Trans. Talcott Parsons London: Unwin Hyman.

Weiser, Mark, & Seely Brown, John (1997) 'The Coming Age of Calm Technology' in Peter J. Denning and Robert M. Metcalfe, Eds. *Beyond Calculation: The Next Fifty Years of Computing*. New York: Copernicus.

Whitaker, Brian (2005) 'Politicians least trusted people' *Guardian Weekly* September 23–29.

Whitrow, G.J (1989) *What is time?* London: Thames and Hudson.

Wiener, Norbert (1947) *Cybernetics: Control and Communication in the Animal and the Machine* Cambridge, Mass.: Technology Press.

Wilkinson, Iain (1999) 'Where is the Novelty in our Current 'Age of Anxiety'? *European Journal of Social Theory*, Vol. 2, No. 4, 445–467.

Williams, Raymond (1974) *Television: Technology and Cultural Form* London: Fontana.

—— (1979) *Politics and Letters* London: New Left Books.

Wilson, Scott (2003) 'Punctuality Drive' *Washington Post*, November 3rd.

Winokur, Mark (2003) 'The Ambiguous Panopticon' ctheory.net http://www.ctheory.net/text_file.asp?pick=371.
Wolin, Sheldon (1997) 'What Time is it?' *Theory and Event* 1.1 http://muse.jhu.edu/journals/theory_and_event/v001/1.1wolin.html.
Xinhua News Agency (2005) http://news.xinhuanet.com/english/2005-06/08/content_3057842.htm.
Young, J. (1993) *Global Network: Computers in a Sustainable Society*, Washington, DC: Worldwatch Institute.
Zakaria, Fareed (2008) *The Post-American World* New York: W.W. Norton.
Zittrain, Jonathan (2008) *The Future of the Internet—And How to Stop It* London: Allen Lane.

INDEX

www.ingramcontent.com/pod-product-compliance
Lightning Source LLC
LaVergne TN
LVHW080847300926
849162LV00003B/361
* 9 7 8 9 0 0 4 1 7 5 9 0 7 *